AF323132

Springer Proceedings in Physics 50

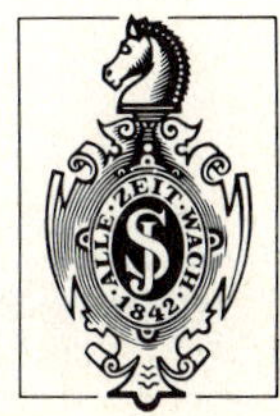

Springer Proceedings in Physics

Managing Editor: H.K.V. Lotsch

Volumes 1–29 are listed on the back inside cover

Magnetic Properties of Low-Dimensional Systems II

New Developments

Proceedings of the Second Workshop,
San Luis Potosí, Mexico, May 23–26, 1989

Editors: L. M. Falicov, F. Mejía-Lira,
and J. L. Morán-López

With 119 Figures

Springer-Verlag Berlin Heidelberg New York
London Paris Tokyo Hong Kong

Professor Leopoldo M. Falicov
Department of Physics, University of California, Berkeley, CA 94720, USA

Professor Francisco Mejía-Lira
Professor José L. Morán-López
Instituto de Física, Universidad Autónoma de San Luis Potosí,
78000 San Luis Potosí, S.L.P., Mexico

ISBN 3-540-52353-7 Springer-Verlag Berlin Heidelberg New York
ISBN 0-387-52353-7 Springer-Verlag New York Berlin Heidelberg

Printing: Weihert-Druck GmbH, D-6100 Darmstadt
Binding: J. Schäffer & Co. KG., D-6718 Grünstadt
2154/3150-543210 – Printed on acid-free paper

Preface

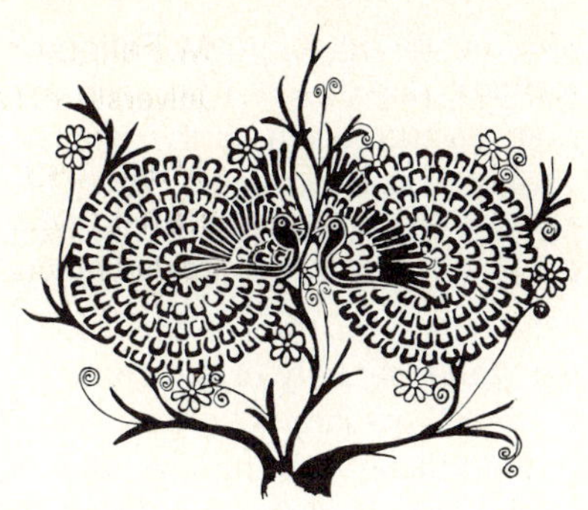

The city of San Luis Potosí is the capital of the Mexican state of the same name. It was originally settled in 1583 and lies 1880 meters above sea level on the central plateau of Mexico. It is characterized by its colonial buildings with multicolored tile decorations, the most famous of which are the cathedral, begun in 1679, and a delightful baroque church, *El Carmen*. The city has played an important role in Mexican history (Francisco I. Madero, for instance, wrote his revolutionary *Plan de San Luis* in 1910 while in prison there), and is an important cultural center of the region. The Universidad Autónoma de San Luis Potosí is located in this town.

In that congenial and historic environment, and making use of more modern and up-to-date facilities, a group of 72 scientists gathered, between 22 and 26 May 1989, to discuss the physical properties of surfaces, thin films, multilayer samples, layered compounds, clusters, chains, linear compounds, disordered materials and other real and idealized condensed-matter systems. The occasion was the Second International Workshop on the Magnetic Properties of Low-Dimensional Systems. This twentieth-century Mexican mini-tradition – the First Workshop was held in 1986 in Taxco, Guerrero, Mexico – was a truly international event supported by a distinguished group of international organizations: the Universidad Autónoma de San Luis Potosí; the Gobierno del Estado de San Luis Potosí (Mexico); the Consejo Nacional de Ciencia y Tecnología (Mexico); the Secretaría de Educación Pública (Mexico); the International Centre for Theoretical Physics (Trieste, Italy); the National Science Foundation – Division of International Programs (United States); the American Physical Society – International Physics Group (United States); and the Centro Latinoamericano de Física (headquarters in Rio de Janeiro, Brazil).

The participants were also representative of many nations: there were 39 registered participants from Mexico, 15 from the United States, 4 from the Federal Republic of Germany, 3 from Switzerland, 2 from Brazil, 2 from Colombia, and 1 each from Argentina, Chile, Costa Rica, England, Japan, Poland and Spain.

The theme of the meeting has become a major driving force in pure and applied science, and a major component of high-technological industrial applications – magnetic recording and information storage and retrieval – which are commanding, at the end of the twentieth century and world-wide, capital investments of several tens of billions of U.S. dollars.

The field, as a research endeavor, remains truly interdisciplinary. It involves many diverse areas of theoretical, experimental and engineering inquiry: sophisticated numerical models and simulations; unusual statistical concepts; the chemistry and metallurgy of transition-metal and rare-earth compounds; the ever-growing technologies of thin-film, overlayer and multilayer preparation and characterization; surface science; the phenomena of diffusion, alloying and segregation; a variety of truly amazing, new electron-microscopy techniques; the chemistry and physics of intercalation; Mössbauer, spin-resonance, optical and electron spectroscopies; the interplay of magnetism with superconductivity; charge- and spin-density waves; macroscopic, mesoscopic and microscopic phenomena. In short the whole panoply of the contemporary science and engineering of materials.

The Workshop was lively and densely programmed. In addition to the remarks in the official opening ceremony and the brief closing remarks by L.M. Falicov, there were, in three and a half days, 29 forty-minute oral presentations, which covered a very large territory. This volume includes 27 of those 29 papers: S.F. Alvarado and H.C. Siegmann chose not to have their lectures included, thus reducing by two-thirds the written contribution of the Swiss delegation.

The book is divided into seven parts, according to subject; it does not follow the sequential order of the presentations. The contributions do not completely cover the ever-expanding field of magnetism in low-dimensional environments. It does not even attempt, for instance, to broach the areas of magnetic recording and of recording engineering. Time and space limitations make the treatment of most covered subjects brief and schematic, to say the least. But, *in toto*, this volume gives a fair account of various important recent developments in a very active field, and a very good sense of vitality of a research area of great scientific and technological importance.

Berkeley, San Luis Potosí
December 1989

L.M. Falicov
F. Mejía-Lira
J.L. Morán-López

Contents

Part 3   Magnetic Superlattices

Part 4   Statistical Mechanics of Low-Dimensional Magnetism

Part 1

Surface and Thin-Film Microstructure: Micromagnetics

I. Surface Magnetic Microstructure

M.R. Scheinfein, J. Unguris, R.J. Celotta, and D.T. Pierce

National Institute of Standards and Technology,
Gaithersburg, MD 20899, USA

The way in which a magnetic solid minimizes its energy through the formation of domain walls is strongly influenced by the presence of the surface. At the surface, a bulk Bloch wall may change into Néel wall in order to reduce the magnetic stray field energy of the ferromagnetic system. We present high spatial resolution images of surface magnetic microstructure obtained by scanning electron microscopy with polarization analysis (SEMPA). Quantitative domain wall profiles at surfaces have been measured for a wide variety of ferromagnetic materials which display asymmetric surface Néel walls for bulk–like thicknesses. We have calculated the magnetic moment configuration at the surface, using bulk magnetic parameters and an iterative micromagnetic energy minimization scheme. The calculated profiles are compared directly with experimental surface magnetization profiles. The surface magnetic microstructure of a surface magnetic topological singularity is observed and an upper limit on the size of the singularity is determined.

I.1. Scanning Electron Microscopy with Polarization Analysis (SEMPA)

Scanning Electron Microscopy with Polarization Analysis (SEMPA), in which the spin polarization of the secondary electrons is measured, is unique in domain imaging techniques using electron microscopy in that the image is directly proportional to the magnetization. In this sense it is like the magneto–optic Kerr effect without the transverse spatial resolution of the technique being limited by diffraction at optical wavelengths. The sampling depth of the SEMPA probe is limited by the escape depth for the spin–polarized secondary electrons, which is estimated to be on the order of 10 to 20 angstroms. This property of SEMPA, coupled with the high transverse spatial resolution attainable with focused electron beams, makes SEMPA ideal for observing surface magnetic microstructure.

In the SEMPA experiment, a beam of high (5 to 60 keV) energy electrons is rastered point–by–point across the surface of a ferromagnetic sample generating spin–polarized secondary electrons at each point [I.1]. The polarized electrons are efficiently extracted from the locale of the sample surface, and focused into an electron–spin polarization analyzer. Most electron–spin polarization analyzers separate electrons of differing spin orientation via the spin–orbit interaction in

Springer Proceedings in Physics, Vol. 50 **Magnetic Properties of Low-Dimensional Systems II**
Editors: L.M. Falicov · F. Mejía-Lira · J.L. Morán-López © Springer-Verlag Berlin, Heidelberg 1990

the scattering of the polarized electron beam with a high–Z target, such as Au. We utilize a new type of spin analyzer, which is exceptionally compact and efficient, developed explicitly for experiments of this type [I.2,I.3]. A map of the surface electron spin–polarization is generated as the focused beam scans the sample surface and the polarization of the emitted secondary electrons is analyzed.

In a 3d ferromagnetic transition metal such as Fe, Co, or Ni, the magnetization M is directly proportional to the net spin density, $n_\uparrow - n_\downarrow$

$$M = -\mu_B \left(n_\uparrow - n_\downarrow \right) \tag{I.1}$$

where $n_\uparrow$ $(n_\downarrow)$ is the number of spins per unit volume parallel (antiparallel) to a particular quantization direction and μ_B is the Bohr magneton. Spin polarized electron spectroscopies such as SEMPA depend on extracting electrons from the solid without loss of any information. One then measures the degree of the spin polarization P of the extracted beam,

$$P = (N_\uparrow - N_\downarrow)/(N_\uparrow + N_\downarrow) \tag{I.2}$$

where $N_\uparrow$ $(N_\downarrow)$ is the number of electrons with spin parallel (antiparallel) to some quantization direction. Low energy secondary electrons are primarily the result of electron–hole pair creation and thus reflect the net spin density of the valence band. The secondary electron spin polarization can be estimated as $P = n_B / n$, where n_B is the agnetic moment per atom and n is the number of valence electrons per atom; one estimates net electron spin–polarization of 28%, 19%, and 5% for Fe, Co, and Ni, respectively.

In our implementation of SEMPA, we simultaneously measure the total secondary intensity, characteristic of topographic and work–function related contrast, and two perpendicular components of the electron spin–components of the in–plane magnetization or one in–plane and the out–of–plane magnetization simultaneously. This is useful when forming maps of the magnetization angle in the surface of the specimen because the two–components of the in–plane magnetization are measured simultaneously, thus eliminating registration problems arising from sample drift.

I.2. Magnetic Microstructure

We would like to emphasize here the profound effect that the surface has in determining the magnetic microstructure. Detailed studies of surface magnetic microstructure are of both fundamental and technological importance. Although there has been intense study of magnetic microstructure for years, there is still considerable uncertainty about relative wall widths, wall profiles and the magnetization distribution about topological singularities. This is due primarily to a lack of a sufficiently high resolution probe to examine the magnetic microstructure experimentally. Fundamental studies will further impact the development of magnetic storage technology, as device densities will ultimately be limited by the spatial extent of the magnetic microstructure.

The principal effect that a surface has on a solid is breaking the translational symmetry in a direction normal to it. This affects the magnetostatic energy of the system in a fundamental way. The presence of the surface forces the magnetization in the sample into the surface plane, thereby minimizing the stray magnetic field energy that would result if the magnetization vector pointed out of a surface. In an infinitely extended ferromagnet, the boundary between anti–parallel domains would be a 180 degree Bloch wall. If the ferromagnet is cut so that the Bloch wall intersects the surface, a large stray magnetic field and an unfavorable energy configuration results. Instead, the magnetization vector rotates within the plane of the surface between anti–parallel domains forming a Néel wall. The internal structure of domain walls between anti–parallel domains varies as a function of the film thickness. For extremely thin films, the magnetization vector rotates in the plane of the film forming Néel walls. For thick, or bulk–like films, the magnetization forms Bloch walls in the interior and Néel walls at the surfaces. For intermediate thicknesses, a number of different wall configurations are possible as first realized by Hubert [I.4] and LaBonte [I.5]. These wall configurations, known as asymmetric Bloch walls, are vortex like structures which have Néel walls at the surfaces, but no well developed wall in the interior. Evidence for this model has been obtained from transmission electron microscopy using Lorentz contrast [I.6] and a newer differential phase contrast teechnique [I.7]. The surface Néel walls predicted by micromagnetic models have been observed by magneto–optic Kerr effect [I.8] and SEMPA [I.1,I.9–I.11]. Lowering the surface magnetic energy of a ferromagnet by formation of surface Néel walls appears to be a widespread phenomenon for relatively soft magnetic materials.

In Fig. I.1, an example of a typical SEMPA image of domains and domain walls found at the (100) surface of an Fe single crystal is shown. The surface of this and other samples described here were prepared by gentle sputtering with a 1 keV Ar ion beam. This sample was re–annealed at 700° C after sputtering. In Fig. I.1a, the horizontal component of the in–plane surface magnetization distribution is shown. The image is 14 μm across. White (black) indicates that the surface magnetization vector points to the right (left). The horizontal component of the magnetization vector is M_x. In Fig. I.1b, the vertical component of the in–plane surface magnetization distribution M_y, is shown. Here, white (black) indicates that the magnetization vector points up (down). This convention will be used throughout. Clearly visible in these images are three large domains. The two largest domains, white and black in Fig. I.1a, are separated by a 180 degree Bloch wall in the bulk and a 180 degree Néel wall at the surface. The Néel wall is the broken horizontal black–white band in Fig. I.1b. There are two topological singularities clearly visible in Fig. I.1b. The first, is at the junc-

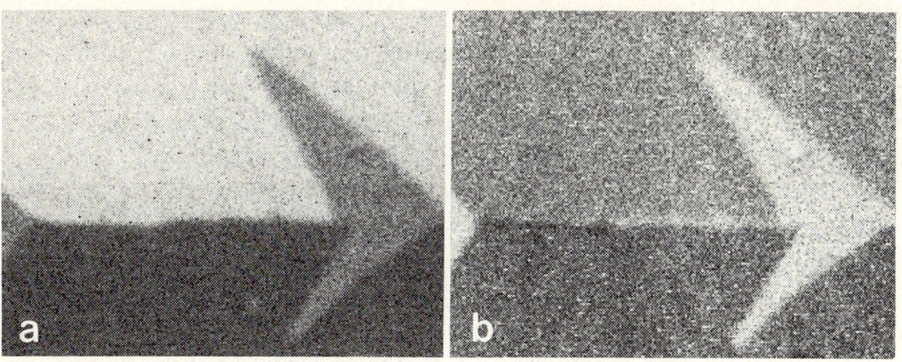

Figure I.1. SEMPA images of an Fe ⟨100⟩ oriented single crystal showing components of the magnetization along the horizontal (a) and vertical (b) directions. The images are 14 micrometers across.

tion of the small white arrow domain on the left and the black surface Néel wall. This singularity has the magnetization circulating in the clockwise sense about it [I.12]. The locus of magnetization vectors about the singularity forms an ellipse. The second singularity is at the junction of the black and white surface Néel walls in the center of Fig. I.1b. This singularity is of another type, where the locus of magnetization vectors forms hyperbolas. The type of magnetic microstructure found on this Fe sample is indicative of that found on bulk samples with in–plane magnetocrystalline anisotropy, namely, large domains separated by surface Néel walls.

Our micromagnetic simulations solve the coupled, non-linear equations derived by Brown [I.13]. In this model, the magnitude of the magnetization vector is constant and equal to the saturation magnetization M_s, while the direction of the magnetization vector may vary. The approach is to minimize the energy of the system subject to the constraint that the magnitude of the magnetization is fixed. The micromagnetic contributions to the total energy of the system are: the nearest neighbor exchange interaction characterized by the bulk exchange coupling constant A; the magneto–crystalline anisotropy energy characterized by the bulk anisotropy contant K_v; the surface magneto–crystalline anisotropy energy characterized by the surface anisotropy constant K_s; and the long range magnetostatic field energy derived from the self–fields generated by the magnetic charges where there is a finite divergence in the magnetic substructure. Our two dimensional simulation follows that of LaBonte [I.5], and the three dimensional simulation follows Schabes and Bertram [I.14]. We divide the ferromagnet into finite sized rods or cubes. The energy and effective magnetic field (the effective magnetic field results from all the energy terms) is calculated for each discretized magnetization element. The angle of each pixel is adjusted so that it points in the direction of the effective magnetic field. As the system relaxes, the magnetization vector at each element will point in the direction of the effective magnetic field, and the local minimum in the energy distribution is reached. Although the energy minimization scheme is computer intensive, taking up to 10 minutes on a supercomputer for a modest grid of 2000 elements, the equilibrium magnetization distribution arises without any ad hoc assumptions which may be model dependent. This phenomenological calculation has as input the magnetic parameters A, M_S, K_v, K_S and the thickness and boundary conditions. The output of the program is the magnetization direction at each pixel and the various contributions to the total energy.

The hard–direction magnetization configuration in a cross–section (the x–z plane) of a 180° wall in a bulk, zero magnetostriction, Co based ferromagnetic glass is shown in Fig. I.2a. The magnetization is assumed to be uniform in the y direction, which is appropiate for the situation discussed here. Note how the strong vertical magnetization of the Bloch wall in the middle of the sample turns over and into the surface plane leading to the surface Néel wall. This results from minimizing the magnetostatic energy. The parameters used in the simulation for the Co based glass are $A = 1 \times 10^{-6}$ erg/cm, M_S=557 emu/cm^3, and $K = 20000$ erg/cm^3. For bulk samples, we determined the critical thickness, such that any further increase in thickness resulted in elongating the interior Bloch wall without modifying the structure of the surface Néel wall profiles.

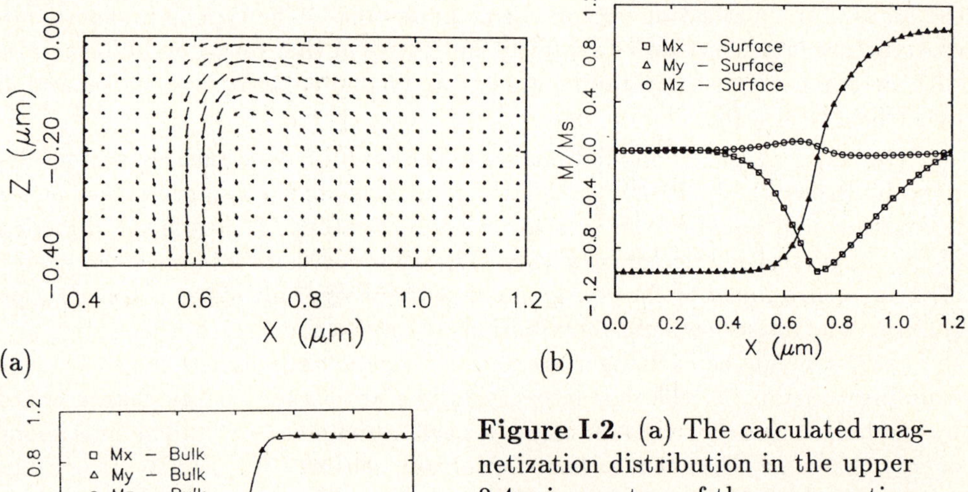

Figure I.2. (a) The calculated magnetization distribution in the upper 0.4 micrometers of the cross section through a Co based ferromagnetic glass sample.
(b) Calculated surface domain wall profiles for each component of the magnetization distribution shown in (a).
(c) Calculated bulk domain wall profiles for each component of the magnetization distribution shown in (a).

For Fe, for example, it was found that the structure of the interior Bloch wall was completely developed for a film thickness of 0.4 μm. The width of the Bloch wall calculated at the center of the 0.4 μm film is identical to that of an infinite crystal calculated using periodic boundary conditions to eliminate surface effects.

The profiles of the surface Néel wall are shown in Fig. I.2b by plotting the components of the surface magnetization. The asymmetry of the surface Néel wall is most easily seen by examining M_x, the surface magnetization component perpendicular to the wall. It clearly falls much more sharply on the left and rises more slowly on the right. The profiles of the bulk Bloch wall, formed by taking a trace through the wall in a direction perpendicular to it, are shown in Fig. I.2c. M_z is the vertical component of the magnetization directed normal to the surface plane. The asymmetry so clearly seen in the surface profiles is no longer evident in bulk. The width of the calculated surface Néel wall is almost twice the width of the bulk Bloch wall.

Note in Fig. I.2a that the center of the Bloch wall in the interior is displaced in the x–direction from the center of the surface Néel wall. This can also be seen by comparing the position of the maximum of M_x in Fig. I.2b with the position of the maximum of M_z in Fig. I.2c. This distance, between the peak of M_z in the interior and the peak of M_x at the surface is Δ.

The measurements in Fig. I.3 are from a zero magnetostriction Co based ferromagnetic glass (Allied 2705M, $Co_{69}Fe_4Ni_1Mo_2B_{12}Si_{12}$). In Fig. I.3a, an image of M_y, one sees parts of two domains with magnetization nearly aligned with the vertical axis. The magnetization within the domain walls, as seen in the M_x image in Fig. I.3b, lies in the horizontal direction, perpendicular to the wall. This demonstrates that, at the surface, the domain wall is a Néel wall with the magnetization rotation occurring in a counter–clockwise direction in the surface plane. These images are about 9 μm across. The point at which the Néel wall changes direction is a magnetic topological singularity on the surface. Note that the black and white segments of the wall are slightly offset from each other. The shift is quite evident in Fig. I.3b; the distance between the center of the black and white walls in the image is 2Δ.

The widths of the surface Néel walls are consistently wider than the corresponding Bloch wall in the bulk. There is no experimental observations of bulk Bloch walls for bulk samples. Transmission electron Lorentz microscopy studies [I.6] of Fe films of up to 0.3 μm thickness show wall widths consistent with our models. However, these walls are in the asymmetric Bloch wall regime, having no well developed Bloch wall in the interior. If reasonable electron transmission through Fe films of thicknesses greater than 0.6 μm could be achieved with, for example, a 1 MeV transmission electron miroscope [I.15], one would be able then to measure a Bloch wall width close to that in bulk. Since the vortex wall structure of the asymmetric Bloch wall is still present in Fe films up to 0.3 μm, it is incorrect to identify the walls measured in this case with bulk Bloch walls.

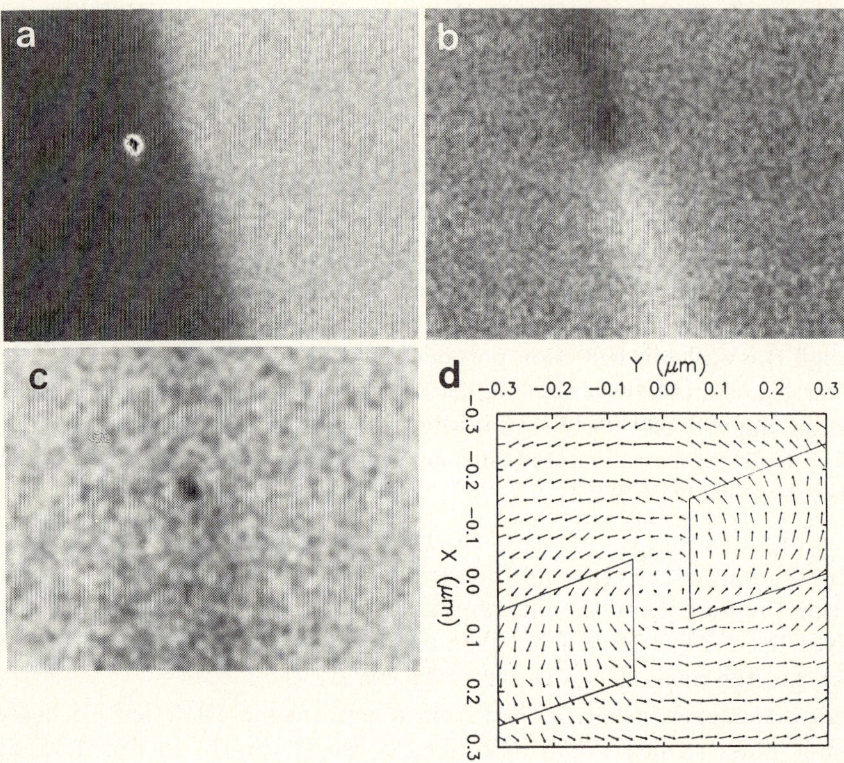

Figure I.3. (a) The vertical component of the magnetization M_y in a Co based ferromagnetic glass. The horizontal extent of the image is 9 μm. (b) The horizontal component of the magnetization recorded simultaneously with (a). (c) The magnitude of the in–plane magnetization. (d) The surface swirl pattern near the singularity at the junction of the two surface Néel walls.

Shown in Fig. I.3c is the magnitude of the in–plane magnetization as calculated from adding the magnetization of Fig. I.3a and I.3b in quadrature. A black dot is clearly visible against the nearly constant background which corresponds to a drop in the magnitude of the in–plane magnetization. This point coincides with the location of the magnetic topological singularity at the surface. The missing magnetization is due to a finite sized probe (60 nm) being convolved with a magnetization distribution which is changing sign on length scales on the order of the probe diameter. When this occurs, the probe samples regions of magnetization with the opposite sign, resulting in a cancellation of some of the signal. In this case, the measured full width at half maximum of the hole in Fig. I.3c is approximately 270 nm in diameter. Shown in Fig. I.3d is the magnetization direction profile in the surface near the singularity derived from our data. The boxed in regions represent the light and dark regions of the surface

Néel walls of Fig. I.3b. The swirl pattern is clearly visible. We attempted to measure the out of plane component of the magnetization near the singularity, but although it was present in some images, it could not be measured repeatedly. With this limitation, we can safely put an upper limit on the radius of the singularity in this Co based ferromagnetic glass at 130 nm. This is still about six times as large as the core of the singularity calculated by Hubert [I.12].

In summary, we find surface Néel wall widths in bulk samples are at least twice those of interior Bloch walls. Our micromagnetic calculations provide a qualitative description of asymmetric surface Néel walls, and predict correctly that the surface wall is offset from the interior Bloch wall at the intersection of a Bloch wall with the surface. SEMPA clearly provides a powerful means for the experimental investigation of surface magnetic microstructure.

The Fe whiskers provided by A. Arrott were grown at Simon Fraser University under an operating grant from the National Science and Engineering Research Council of Canada. This work was supported in part by the Office of Naval Research.

References

I.1 G. G. Hembree. J. Unguris, R. J. Celotta, and D. T. Pierce, Scanning Microscopy Supplement **1**, 229 (1987).

I.2 J. Unguris, D. T. Pierce, and R. J. Celotta, Rev. Sci. Instrum. **57**, 1314 (1986).

I.3 M. Scheinfein, J. Unguris, J. J. McClelland, D. T. Pierce, R. J. Celotta, and M. H. Kelly, Rev. Sci. Instrum. **61**, 1 (1989).

I.4 A. Hubert, Phys. Status Solidi **32**, 519 (1969).

I.5 A. E. LaBonte, J. Appl. Phys. **40**, 2450 (1969).

I.6 T. Suzuki and K. Suzuki, IEEE Trans. Magn. **MAG–13**, 1505 (1977).

I.7 J. N. Chapman, G. R. Morrison, J. P. Jakubovics, and R. A. Taylor, J. Mag. Mag. Mat. **49**, 277 (1985).

I.8 F. Schmidt, W. Rave, and A. Hubert, IEEE Trans. Magn. **MAG–21**, 1596 (1985).

I.9 K. Koike, H. Matsuyama, H. Todokoro, and K. Hayakawa, Scanning Microscopy Supplement **1**, 241 (1987).

I.10 J. Kirschner, Phys. B1 **44**, 227 (1988).

I.11 H. P. Oepen and J. Kirschner, Phys. Rev. Lett. **62**, 819 (1989).

I.12 A. Hubert, Proceedings of ICM 1988 (in press).

I.13 W. F. Brown, Jr., *Micromagnetics* (Wiley, New York, 1963).

I.14 M. E. Schabes, H. N. Betram, J. Appl. Phys. **64**, 1347 (1988).

I.15 J. Sevely, G. Zanchi, Y. Khin, K. Hssein, Scanning Microscopy Supplement **1**, 179 (1987).

II. Theoretical Predictions for Magnetic Hysteresis

R.H. Victora

Eastman Kodak Company, Research Laboratories –Diversified Technologies,
Rochester, NY 14650, USA

Two types of calculations for hysteretic phenomena are described. The first uses a detailed treatment of the relevant interactions, applied to samples of known microstructure, to predict hysteresis loops in unprecedented agreements with experiment. The second includes thermal fluctuations in deducing two scaling relationships applicable to a variety of materials.

II.1. Introduction

Hysteresis is one of the primary manifestations of ferromagnetism, and the consequent experimental and theoretical attention has provided considerable qualitative understanding. Nevertheless, many of the proposed explanations are incomplete or contradictory, and it is clear that the development of quantitative theories is of crucial importance for further progress in this field. However, quantitatively accurate predictions from measurable microscopic input are quite difficult for a variety of experimental and theoretical reasons [II.1]. Accurate characterization of grain boundaries, segregated areas, dislocations, and other irregularities is problematic and the results can vary widely from sample to sample. Many phenomena depend on microscopic nucleation or pinning effects in competition with macroscopic forces. Thermal fluctuations can lead to strong time and temperature dependences. Finally, the magnetostatic interactions and, to a lesser extent, quantum mechanical exchange, make this a many–body problem.

Despite these difficulties, several prominent advances have occurred. In 1948, Stoner and Wohlfarth [II.2] introduced a particulate theory which assumed coherent rotation of the particle magnetization and neglected interparticle interactions: The approach provided order–of–magnitude estimates for coercivities and is still frequently used to gain approximate understanding. The problem of incoherent rotations has been subsequently examined in an idealized form by several authors [II.3]; interactions were successfully treated by Hughes [II.4] in his calculations for CoP thin films. Thermal fluctuations were examined by Néel [II.5] who introduced a simplified model in which the energy barrier is proportional to $(\Delta H)^2$ where ΔH equals the difference of the applied field and the nonthermally assisted switching field. These papers, while not achieving a demonstrated quantitative accuracy, have provided a solid foundation for further theoretical development.

Springer Proceedings in Physics, Vol. 50   **Magnetic Properties of Low-Dimensional Systems II**   11
Editors: L.M. Falicov · F. Mejía-Lira · J.L. Morán-López © Springer-Verlag Berlin, Heidelberg 1990

Reported here are two recent calculations which use micromagnetic principles to make quantitative predictions for measurable macroscopic properties.In the first section, hysteresis loops for two CoNi thin films are predicted [II.6], using only microstructural properties as input, which demonstrate close agreement with experiment. The second section details the derivation [II.7] of a scaling relationship between switching fields and measurement time; exponents are predicted for two phenomena that should be measurable. The concluding section includes a description of two areas for future research.

II.2. Hysteresis Loops in CoNi Thin Films

In this section, attention will be focused on two CoNi thin films formed [II.8] by oblique evaporation and consisting of long columnar grains inclined at an angle to the plane of the film (see Fig. II.1). Both samples have thicknesses of approximately 1500 Å and angles of inclination of 50° and 60° to the surface normal. Each grain is approximately 200 Å in diameter. The grain boundaries have a significantly different texture than do the grains under electron microscopy, and it is probable that they are partly vacuum. Diffraction studies show the **c**–axis of the *hcp* structure to have a wide directional dispersion, which probably tends along the columnar direction.

For calculational purposes the thin film was divided into cells that contain a number of complete grains (see Fig. II.1). These cells were linked by periodic boundary conditions, and several sizes ranging from 3×3 to 12×10 grains were used. A variety of grain–boundary patterns, including one taken directly from an electron micrograph, were used; results were found to be insensitive to the precise pattern provided that average grain size, inclination angle, and the proportion of grain boundary were correct. Each grain was then divided along its major axis into segments such that each segment was approximately cubic.

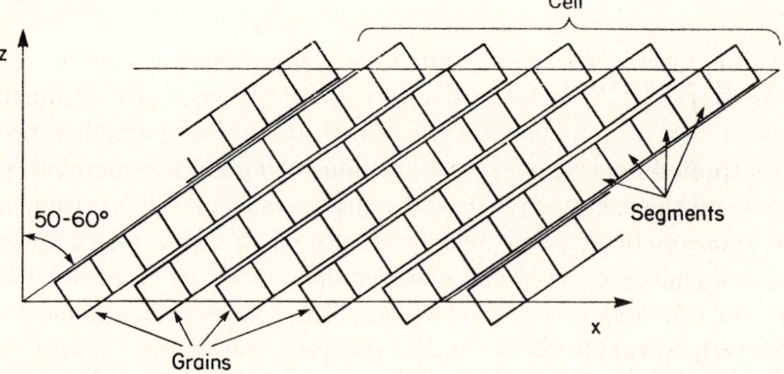

Figure II.1. Simplified description of how cell, grain, and segment boundaries are defined for this calculation. As described in the text, actual grain boundaries used in this calculation are substantially more complex.

Magnetostatic interactions between all segments are included. For closely neighboring segments (typically within 1500 Å), the interaction tensor for each pair of segments is calculated [II.6] by a finite–element integration. For more distant particles, the interaction is included through a mean–field approximation.

The *quantum–mechanical exchange* parameter of bulk $Co_{80}Ni_{20}$ may be approximated by appropriately averaging the values for pure Co and Ni [II.9]. This value provides an upper bound for the true value within a grain, which is known to contain dislocations and probably segregations. It is found that calculation results are insensitive to values of the exchange parameter between 1/4 and the full bulk value. Thus the calculation is independent of reasonable choice for the exchange parameter and 1/2 the bulk value is used: $AM_0^2 = 1.32 \times 10^{-6}$ ergs/cm. Quantum–mechanical exchange between grains is neglected owing to its expected small value.

The total *anisotropy* of the thin film may be calculated from torque magnetometry. However, the shape contribution to this due to demagnetization fields may also be calculated for each particular arrangement of grains. The difference is, of course, the crystalline anisotropy, which is assumed to be evenly divided between segments and becomes an effective field in the calculations. The 60° sample is found to have negligible net crystalline anisotropy, whereas the 50° grain has crystalline anisotropy $K = 5 \times 10^5$ ergs/cm³ within grains. This may be compared to the much larger values calculated for the shape anisotropy: approximately 1.5×10^6 ergs/cm³ (directed along the columnar axis) for both samples.

The *Landau–Lifshitz–Gilbert equations* are believed to give an adequate description for the time development of the magnetic moment in the case of fixed magnitude [II.10,II.11]. This requirement is fulfilled for our small exchange–coupled segments and thus each segment is assigned a magnetization vector $\vec{M}$, which moves according to:

$$\frac{d\vec{M}}{dt} = \frac{-\gamma}{1 + \alpha^2} \vec{M} \times \vec{H} - \frac{\alpha\gamma}{1 + \alpha^2} \hat{M} \times (\vec{M} \times \vec{H}).$$

Here $\vec{H}$ includes the applied magnetic field, the magnetostatic interaction fields, the anisotropy fields, and the exchange field. To a large extent, the gyromagnetic ratio γ and damping parameter α are arbitrary for the nearly static phenomena described here; I have chosen the free electron values for $\gamma = 1.76 \times 10^7$ (G sec)$^{-1}$ and a value $\alpha = 1$ for numerical convenience. The temporal integration is conducted using a fourth–order Runge–Kutta [II.12] scheme with $\Delta t = 1 \times 10^{-11}$ sec.

Microscopic *domain nucleation* sites play a major role in magnetization reversal for many materials. Inclusion of these various sites, many of them of unknown nature, would be quite difficult in a calculation of this scale. Instead,

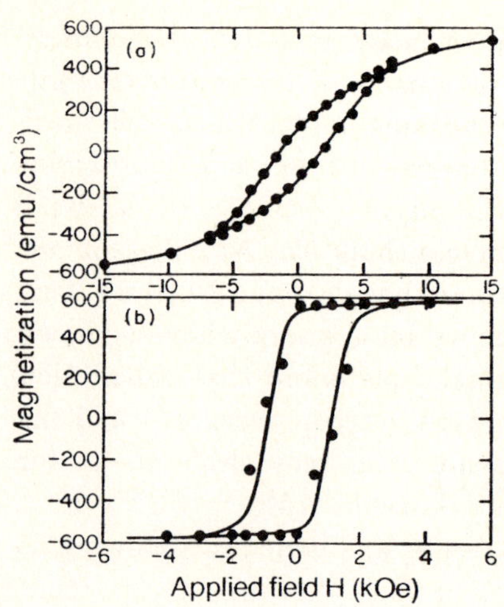

Figure II.2. Experimental (solid line) and theoretical (dots) hysteresis loops. (a) Perpendicular loop for the 60° sample. (b) Longitudinal loop for the 60° sample. (Experimental data taken by L. S. Meichle).

our treatment is based on the expectation that the nucleated domain can easily grow until it reaches a grain boundary or the magnetostatic shape effects of the grain become large. Thus, to simulate the effects of domain nucleation within this calculation, a single segment within each grain is flipped along the direction of the applied field and this reverse domain is allowed to grow or decay without further intervention. Once equilibrium has been reached and a new external field applied, the process is repeated.

Hysteresis loops for the perpendicular and longitudinal (along the inplane projection of the columnar direction) directions of the 60° film are shown in Fig. II.2. The agreement with experiment is excellent; in fact, given the approximately 5% level of error in several experimental input variables, the agreement for the perpendicular loop may be regarded as fortuitous. Physically it is found that domain nucleation and propagation cause longitudinal hysteresis, with individual grains reversing at different fields and apparently stabilizing their neighbors. On the other hand, domains nucleated in the calculation for the perpendicular loops are typically found to rapidly decay and in no case are they found to have substantial influence on the shape of the loop. In other words, coherent rotation is the dominant effect in perpendicular hysteresis. Finally, hysteresis loops for the 50° sample have also been calculated and shown to have the same excellent level of agreement: perpendicular coercivity is 2100 Oe for both theory and experiment, longitudinal coercivities are 1350 Oe (experiment) *vs.* 1500 Oe (theory).

II.3. Predicted Time Dependence of the Switching Field

The preceding section demonstrated that rigorous theoretical predictions for hysteresis can be made for a sample that is well characterized and in which detailed understanding of the thermally activated processes is not necessary.

Frequently, the opposite is the case: for these situations, scaling relationships relating two or more physical variables can be very useful. This section will provide a derivation of such a relationship between measurement time and switching field in the case where the height of the energy barrier to thermal fluctuations is much greater than $k_B T$. This corresponds to the physically and technologically interesting case of measurement frequency much less than the attempt frequency and is the situation usually considered. It implies that, to a high degree of accuracy, a Maxwell–Boltzmann distribution occurs in the neighborhood of the switching point [II.13]. This means [II.5] that the probability of switching is given by

$$P = A\tau \exp(-\Delta/k_B T).$$

Here, τ = measurement time, ΔE = energy barrier, and A is the attempt frequency which is approximately 10^9 Hz. Actually, A depends on the applied field; however, in the conditions considered here ($\Delta E \gg k_B T$), this contribution to τ is extremely weak in comparison to the ΔE contribution and will be neglected.

We begin the derivation by considering the sample, consisting of numerous small areas of magnetization all interacting with each other both magnetostatically and through quantum–mechanical exchange, to be subject to applied field H with the nearest high frequency (i.e., without thermal fluctuations) switching field labelled H_0 . The energy is Taylor expanded about the state at $H = H_0$ in the applied field $\Delta H = H_0 - H$ and the numerous variables $\Delta\theta_i = \theta - \theta_0$ representing the magnetization degrees of freedom. The expansion is carried to third order. The first order terms, except for the ΔH dependence, are all equal to zero because the system is in a local minimum. The matrix of second derivatives is diagonalized to obtain the normal modes of the system. Henceforth, the problem shall be addressed in terms of these normal modes:

$$\Delta\beta_i = \sum_j A_{ij}\Delta\theta_j.$$

Note that these normal modes do not usually represent the motion of a single grain's magnetization; typically, though, they will be localized.

The second derivative of the energy with respect to the normal mode (or modes) corresponding to $H = H_0$ will be zero. This is a consequence of the definition of a switching point: the point where a neighboring minimum and maximum meet. The assumption will be made that the third derivative with respect to the normal mode does not vanish. There are certain highly symmetric situations where this is not true; for example, the axially symmetric situation examined in the Néel model [II.5] exhibits a third derivative equal to zero. However, the usual physical situation is not symmetric and the third derivative does exist. The same physical intuition also prompts us to ignore the possibility of accidental degeneracies and thus simplify that analysis by considering a single switching mode.

The energy expansion may now be written:

$$E = E_0 - \left.\frac{\partial E}{\partial H}\right|_0 \Delta H + \frac{1}{2}\sum_i \left.\frac{\partial^2 E}{\partial \beta_i^2}\right|_0 (\Delta\beta_i)^2 - \sum_i \left.\frac{\partial^2 E}{\partial \beta_i \partial H}\right|_0 (\Delta\beta_i)(\Delta H)$$

$$+ \frac{1}{6}\sum_{i,j,k} \left.\frac{\partial^2 E}{\partial \beta_i \partial \beta_j \partial \beta_k}\right|_0 (\Delta\beta_i)(\Delta\beta_j)(\Delta\beta_k)$$

$$- \frac{1}{2}\sum_{i,j} \left.\frac{\partial^3 E}{\partial H \partial \beta_i \partial \beta_j}\right|_0 (\Delta H)(\Delta\beta_i)(\Delta\beta_j) + O(\Delta H)^2 .$$

$$(II.1)$$

Here we have anticipated our answer and not listed the higher order contributions $(\Delta H)^2$. The next step is to solve for the $\Delta\beta_i$ by requiring that the system be at an energy extremum. Therefore:

$$\frac{\partial E}{\partial \beta_i} = \left.\frac{\partial^2 E}{\partial \beta_i^2}\right|_0 (\Delta\beta_i) - \left.\frac{\partial^2 E}{\partial \beta_i \partial H}\right|_0 (\Delta H) + \frac{1}{2}\sum_{j,k} \left.\frac{\partial^3 E}{\partial \beta_i \partial \beta_j \partial \beta_k}\right|_0 (\Delta\beta_j)(\Delta\beta_k)$$

$$- \sum_j \left.\frac{\partial^3 E}{\partial H \partial \beta_i \partial \beta_j}\right|_0 (\Delta H)(\Delta\beta_j) = 0 . \qquad (II.2)$$

We know that, aside from the switching mode $\Delta\beta_s$, the values of $\Delta\beta_i$ represent single–valued extensions about $\Delta\beta_i = 0$ since we have required that H_0 be the nearest switching point. This means that the second order terms in $\Delta\beta_i$ may be neglected without affecting the lowest order result. Therefore:

$$\Delta\beta_i = \frac{-\dfrac{1}{2}\left.\dfrac{\partial^3 E}{\partial \beta_i \partial \beta_s^2}\right|_0 (\Delta\beta_s)^2 + \left.\dfrac{\partial^2 E}{\partial H \partial \beta_i}\right|_0 \Delta H + \sum_{j \neq i} \left.\dfrac{\partial^3 E}{\partial H \partial \beta_i \partial \beta_j}\right|_0 (\Delta H)(\Delta\beta_j)}{\left.\dfrac{\partial^2 E}{\partial \beta_i^2}\right|_0 - \left.\dfrac{\partial^3 E}{\partial H \partial \beta_i^2}\right|_0 \Delta H} .$$

Thus, $\Delta\beta_i$ is of order ΔH or $(\Delta\beta_s)^2$ for $i \neq s$. Insertion into (II.1) yields, for the case of $\Delta\beta_i = O(\Delta H)$, a contribution to the energy of order at least $(\Delta H)^2$. Similarly, the case $\Delta\beta_i = O(\Delta\beta_s)^2$ produces terms of order $(\Delta\beta_s)^4$ and $(\Delta H)(\Delta\beta_s)^2$, which are of higer order than other nonzero terms already in (II.1). Thus, the nonswitching modes of the system either make no contribution to the lowest order term or contribute at order $(\Delta H)^2$.

Eq. (II.2) may now be solved for the switching mode to yield:

$$\Delta \beta_S = \pm \sqrt{2\Delta H \frac{\partial^2 E}{\partial \beta_S \partial H}\bigg|_0 \bigg/ \frac{\partial^3 E}{\partial \beta_S^3}\bigg|_0} + O(\Delta H).$$

This corresponds to the values of $\Delta \beta_S$ at the local energy minimum and at the local saddle which the magnetization must cross to switch. The height of this barrier is given by inserting the values of $\Delta \beta_S$ into (II.1) to yield:

$$\Delta E = (\Delta H)^{3/2} \sqrt{2 \left(\frac{\partial^2 E}{\partial \beta_S \partial H}\bigg|_0\right)^3 \cdot 4 \bigg/ \left(9 \frac{\partial^3 E}{\partial \beta_S^3}\bigg|_0\right)}. \tag{II.3}$$

Thus, one concludes:

$$k_B T \ln(2A\tau) = C(\Delta H)^{2/3}.$$

where C is the scaling constant contained within the square root of (II.3).

This result affects the predictions of a number of previous papers which based their analysis upon the $(\Delta H)^2$ dependence described in Néel's simple model. Most obviously, the effect of time in relaxing a system changes from a $[\ln(2A\tau)]^{1/2}$ dependence to a $[\ln(2A\tau)]^{2/3}$ dependence. Less obviously, the energy barrier preventing fluctuation–induced switching will tend to be larger than the Néel result for a given ΔH. This will lead workers to claim lower effective switching volumes based on their measurements. A particularly interesting change will be found in the prediction of Charap [II.14] for the maximum rate of magnetization decay normalized to the saturation magnetization. Using Néel's model, he obtained approximately a $T^{1/2}$ variation by assuming a Lorentzian switching field distribution; the new result would be $T^{2/3}$.

Experimental verification of these predictions would obviously be desirable. It is easy to show that the experimental measurements of Oseroff *et al.* [II.15] for coercivity as a function of time (their Fig. 4) fit a $(\Delta H)^{3/2}$ dependence very well. (The comparison is made for the five swept field points since this avoids their discrepancy as they move to a "jumped" field). However, a $(\Delta H)^2$ dependence cannot be ruled out. An accurate determination of the exponent by this method may require several measurements at still higher frequency. Alternatively, it may be possible to exploit very low temperature measurements of the maximum magnetization decay rate to test the $T^{2/3}$ dependence discussed in the previous paragraph. Here, special care must be taken that the applied field and temperature are very constant. This measurement was attempted by Tobin *et al.* [II.16]; unfortunately, their data are too scattered at the lowest temperatures to realiably fit an exponent, and a fit to higher temperatures shows a strong dependence of the exponent on the cutoff temperature.

II.4. Conclusion

Two theoretical attempts to predict hysteretic phenomena have been described. The first involved a detailed calculation, for two CoNi thin film samples, which included domain nucleation, incoherent rotation of grain magnetization, and complete magnetostatic interactions. Input to these calculations came from a detailed knowledge of the sample microstructure drawn from sources such as electron microscopy. The accuracy of the calculational method and experimental input permitted quantitative predictions for hysteresis in unprecedented agreement with experiment: the longitudinal and perpendicular loops matched experiments to within 10%.

The second calculation featured a derivation of a scaling relationship between measurement time and switching field. The predicted $(\Delta H)^{3/2}$ dependence differs from the commonly used Néel model's $(\Delta H)^2$ dependence: this is because the Néel model is derived for a special case of high symmetry which is usually not applicable. The new scaling relationship implies that the maximum magnetic viscosity at low temperature should exhibit a $T^{2/3}$ dependence and that phenomenologically obtained switching volumes are actually smaller than previously thought. Several of these predictions offer opportunities for experimental verification.

Both calculations suggest directions for future research. An identification of the domain nucleation mechanism postulated in Section II.2 should yield even more accurate predictions of hysteresis and clues towards improved materials. Some initial progress [II.17] has been made for the case of barium ferrite particles; however, a definitive resolution was not achieved. Another important direction is calculation of the easily measured scaling constant C mentioned in Section II.3. This constant will depend on the nucleation mechanism and may be sensitive to fluctuations in the atomic moments. It is hoped that progress in these directions, together with calculations such as those presented here, will lead to a more fundamental understanding of magnetic hysteresis.

I would like to thank A. K. Agarwala, C. F. Brucker, C. Byun, J. S. Meichle, S. B. Oseroff, J. P. Peng, S. Schultz, V. N. Tobin, and W. E. Yetter for providing experimental data and/or other valuable assistence.

References

II.1 For a more complete description, see: W. F. Brown, *Micromagnetics* (Wiley, New York, 1963).

II.2 E. C. Stoner and E. P. Wohlfarth, Philos. Trans. R. Soc. London, Ser. A **240**, 599 (1948).

II.3 See, for example: E. H. Frei, S. Shtrikman, and D. Treves, Phys. Rev. 106, 446 (1957); E. Della Torre, IEEE Trans. Magn. **21**, 1423 (1985).

II.4 G. F. Hughes, J. Appl. Phys. **54**, 5306 (1983).

II.5 L. Néel, Ann. Geophys. **5**, 99 (1949); Adv. Phys. **4**, 191 (1955).

II.6 R. H. Victora, Phys. Rev. Lett. **58**, 1788 (1987); J. Appl. Phys. **62**, 4220 (1987).

II.7 R. H. Victora, Phys. Rev. Lett. **63**, 457 (1989).

II.8 J. D. Gau and W. E. Yetter, J. Appl. Phys. **61**, 3807 (1987).

II.9 G. Shirane, V. J. Minkiewicz, and R. Nathans, J. Appl. Phys. **39**, 383 (1968).

II.10 M. Sparks, *Ferromagnetic Relaxation Theory* (McGraw-Hill, New York, 1964).

II.11 A. P. Malozemoff and J. C. Slonczewski, *Magnetic Domain Walls in Bubble Materials* (Academic Press, New York, 1979), p. 30–39.

II.12 G. F. Simmons, *Differential Equations* (McGraw-Hill, New York, 1972).

II.13 S. Chandrasekhar, Rev. Mod. Phys. **15**, 1 (1943), especially pp. 63–68.

II.14 S. H. Charap, J. Appl. Phys. **63**, 2054 (1988).

II.15 S. B. Oseroff, D. Franks, V. M. Tobin, and S. Schultz, IEEE Trans. Magn. **23**, 2871 (1987).

II.16 V. M. Tobin, S. Schultz, C. H. Chan, and S. B. Oseroff, IEEE Trans. Magn. **24**, 2880 (1988).

II.17 R. H. Victora, J. Appl. Phys. **63**, 3423 (1988).

Part 2

Atomic-Scale Structure of Magnetic Surfaces and Overlayers

III. Surface Magneto-Optical Studies of Ultrathin Ferromagnetic Films

C. Liu and S.D. Bader

Materials Science Division, Argonne National Laboratory,
Argonne, IL 60439, USA

The surface magneto–optic Kerr effect is used to study the existence of mono-layer magnetism and the surface magnetic anisotropy of epitaxial films of Fe deposited on Cu(100), Ru(0001), and Pd(100). The critical behavior of Fe/Pd(100) is also characterized and found to be in agreement with two–dimensional Ising-model results.

III.1. Introduction

The surface magneto–optic Kerr effect (SMOKE) was introduced in 1985 to study ultrathin ferromagnetic films extending into the monolayer and submono-layer regime [III.1]. Since that time SMOKE has contributed significantly to the great advances that have taken place in the field of surface magnetism [III.2–III.4]. In the present work the most recent activities in our laboratory are surveyed involving the growth of Fe on Cu(100), Ru(0001), and Pd(100). These studies provide insights into three topics of interest: (i) the quest for monolayer magnetism; (ii) the nature of the surface magnetic anisotropy; and (iii) critical phenomena in two-dimensional magnetic systems. In the following we outline the experimental aspects of the SMOKE technique, the structural aspects of the systems, and the results of interest.

III.2. Experimental Technique

The films are grown in 10^{-11} Torr ultrahigh vacuum (UHV). In–situ SMOKE characterizations utilize crossed magnetic fields that are oriented in-plane and normal to the film plane, to sequentially monitor the corresponding magnetization components. The two configurations of the magnetic field are referred to as the longitudinal and polar Kerr–effect geometries, respectively. The electromagnets are located in UHV, while the optical source, analyzer, and the photodiode detector are separated from UHV by a window. The source is a p–polarized He–Ne laser. The reflected light has its polarization rotated due to the magneto–optic interaction. The analyzer is a crystal prism polarizer that is nearly crossed with the incident polarization. Reversing the direction of the magnetization reverses the rotation of the reflected light and changes the intensity at the detector. Sweeping the field provides magnetic hysteresis curves, whose height, referred to as the Kerr intensity, is proportional to the

Springer Proceedings in Physics, Vol. 50 **Magnetic Properties of Low-Dimensional Systems II**
Editors: L.M. Falicov · F. Mejía-Lira · J.L. Morán-López © Springer-Verlag Berlin, Heidelberg 1990

magnetization M. If the easy axis is in–plane along the applied–field direction the longitudinal SMOKE signal yields a square loop and the polar signal yields no hysteresis. If the easy axis is vertical the polar loop is square and the longitudinal signal is absent.

III.3. Structural Characterizations

III.3.1 Epitaxy

Low–energy electron diffraction (LEED) is used to monitor the epitaxy. The Fe films all grow epitaxially in the range reported, based on the *fcc* or γ–Fe structure, and take on the substrate in–plane lattice spacing. The interplanar spacings are characterized by layer–dependent tetragonal distortions due to both interfacial strain and surface relaxation. The disortions are approximately -2% and $+2\%$, respectively, for Fe/Cu(100), based on quantitative LEED studies [III.5]. The disortions are anticipated to be more pronounced for the Ru and Pd cases than for Cu because of the in–plane lattice mismatch. The *fcc* Fe(111) orientation grows on Ru(0001) with the expanded Ru intralayer spacing of 2.70 Å *vs.* 2.54 Å for the equilibrium γ–Fe value, which is obtained by extrapolation from above 910° C where bulk–Fe is stable. The Fe/Pd(100) structure is expected to be body centered tetragonal (*bct*) based on recent quantitative LEED studies for the related system Mn/Pd(100) [III.6]. The *bct* phase is derived from distortion of the *fcc*–Fe structure. The Pd(100) intraplanar distances are 2.75 Å. The tetragonal interplanar distortions inherent in epitaxial films have important consequences for the magnetic anisotropy because the magnetocrystalline contribution could favor vertical easy axes of magnetization [III.7]. This point is discussed with the anisotropy results below.

III.3.2 Growth Mode

Auger electron spectroscopy is used to monitor the surface cleanliness and the films growth mode. The standard procedure is to measure the substrate and Fe Auger signals during growth, and then to compare them to growth–model predictions [III.7,III.8]. The general interface is that the films grow predominantly in a layer–by–layer fashion. Advanced structural techniques, such as RHEED and photoelectron diffraction, have helped to refine our growth techniques [III.9]. From these studies we learn that homoepitaxial growth at elevated temperatures prior to Fe deposition atomically smoothes the substrate surface, and that subsequent low–temperature Fe deposition promotes discrete interfaces. Interfacial mixing for Fe/Cu(100) has dramatic consequences, in that it can stabilize the antiferromagnetic state of γ–Fe [III.10].

III.4. Results and Discussion

III.4.1 Monolayer Magnetism

Figs. III.1–III.3 summarize our magnetic characterizations of Fe on Cu, Ru, and
Pd respectively. The existence of monolayer (ML) magnetism is related to two
issues: the ability to create monolayers, and the intrinsic electronic structure
of the monolayer. For Fe/Cu(100) a ferromagnetic response is not detected at
the monolayer level because of the tendency to intermix [III.2]. The problem
is related to the high surface free energy of Fe relative to that of Cu. For this
reason growth on Ru was initiated. However, Fig. III.2 shows that monolayer
levels of Fe on Ru(0001) also lack a ferromagnetic signature. We related this
to the hybridization of the Fe and Ru d–electrons across the interface. For dilute Fe impurities in a Ru host it is well known that hybridization suppresses
moment formation [III.11]. For Fe on Pd(100) Fig. III.3 shows that even submonolayer magnetism is detected. The hybridization in this case is expected not
only preserve the Fe moment, but also to induce moments on neighboring Pd
sites. Recall that dilute alloys of Fe in Pd hosts are classic giant–moment systems [III.12]. Thus, the interactions with the substrate can yield compositional
gradients and/or electronic structural modifications that govern the magnetic
state of the film.

III.4.2 Surface Magnetic Anisotropy

Figs. III.1–III.3 also summarize the anisotropy results. For Fe/Cu(100) the
region of perpendicular spin orientations is delineated as a function of growth
temperature and films thickness. The Ru and Pd results are at fixed growth-

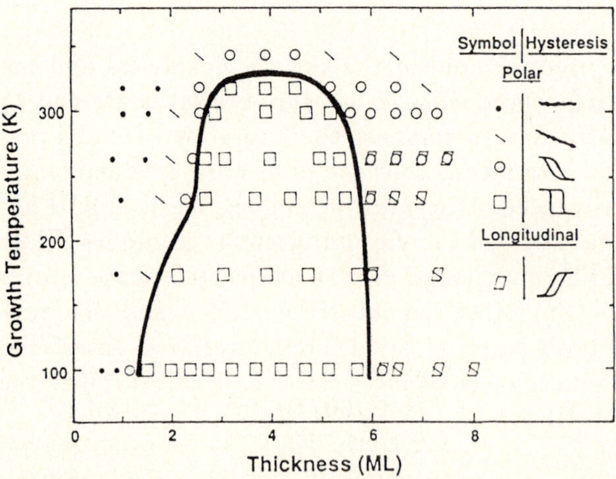

Figure III.1. Anisotropy diagram for Fe/Cu(100). Note the region of stability
of perpendicular easy axes with square hysteresis loops in the polar SMOKE
signal.

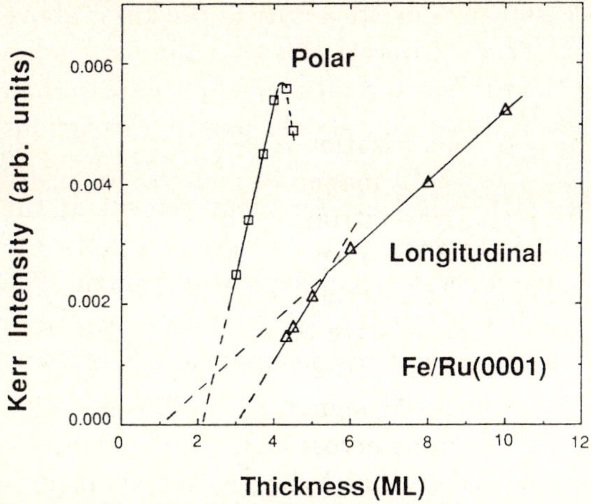

Figure III.2. Anisotropy diagram of Fe/Ru(0001) grown at room temperature plotted as the height of the hysteresis curve in the remanent state. Note the critical thickness where the magnetization reorients in–plane; the longitudinal SMOKE signals become finite and the polar signal vanishes. Note also the lack of ferromagnetism signature at the monolayer level based on null measurements and dashed–line extrapolations.

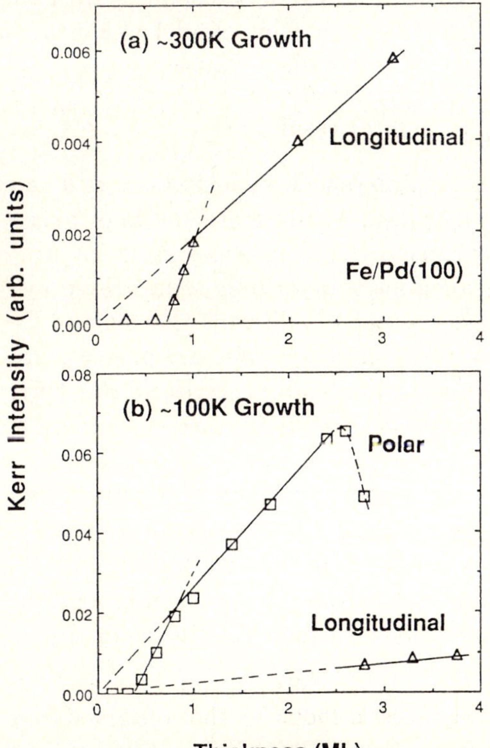

Figure III.3. Anisotropy diagrams for Fe/Pd(100) for (a) 300 K and (b) 100 K growth. Note the critical thickness transition in (b), and the existence of submonolayer magnetism.

temperatures. Note that in all three cases there exists a critical thickness above which the easy axis reorients in–plane. This provides a clue that the perpendicular anisotropy originates more from the surface and interfacial properties than from the geometric distortions discussed above that should persist throughout the thickness range.

The anisotropy diagrams of Figs. III.1–III.3 are for measurement at the growth temperature. The vertical state, however, can be retained above the high–temperature boundaries by raising temperature subsequent to growth. The 1.2 ML Fe/Pd(100) film grown at $\sim$100 K could be taken above the 270 K boundary (not shown in Fig. III.3) and up to the Curie temperature T without reorienting.

III.4.3 Critical Behavior

Dürr *et al.* [III.13] recently extracted the critical magnetization exponent β_C for Fe/Au(100), and reviewed previous work in the field. There are numerous motivations to purse the issue further. Firstly, we have found that Fe/Pd(100) has an advantage of thermal stability and reversibility, while Fe/Au(100) can be susceptible to segregation and intermixing problems [III.8]. Secondly, Fe/Pd(100) has the additional advantage of permitting studies of vertical as well as in–plane spin orientations, while the magnetization of Fe/Au(100) is only in–plane. Finally, the β–value of 0.22$\pm$0.05 reported for Fe/Au(100) disagrees with the expected two–dimensional Ising β_C=1/8=0.125. An Ising–model exponent is expected even though the films are physical realizations of Heisenberg systems [III.14] because even arbitrarily small anisotropy causes the two–dimensional Heisenberg system to yield an Ising–like transition.

Fig. III.4 shows the M–$vs.$–T curves for different film thicknesses and spin orientations, including warming and cooling data to illustrate the high degree of reversibility of the transitions. M represents the normalized Kerr intensity measured in the remanent state (H=0) for films that exhibit square–loop hysteresis curves. Thus, no external fields are present to add tails that extend the transitions to higher temperatures. However, significant tails are present, presumably due to the interactions with the highly polarizable interfacial Pd. Dürr, *et al.* [III.13] outlined methodology for extracting β in the presence of high–temperature tailing superimposed on the power–law behavior. We follow their procedure and parametrize T_C to extend the range of linearity of the traditional log–log plots used to extract β, where $M \propto (1 - T/T_C)^\beta$. The results are shown in Fig. III.5 and the T_C values are plotted in Fig. III.6. The 1.2 ML film has the densest data set, and, therefore, was subjected to the mosy thorough analysis, which yields β=0.127$\pm$0.004. Thus, agreement with the two–dimensional Ising result of β=0.125 is achieved.

The trend in the T_C–values in Fig. III.6 is compared with that obtained from Monte Carlo calculations of Binder and Hohenberg [III.15] for two–dimensional

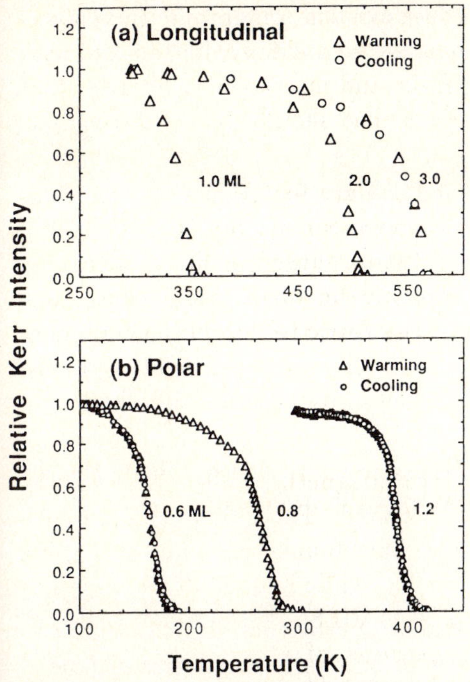

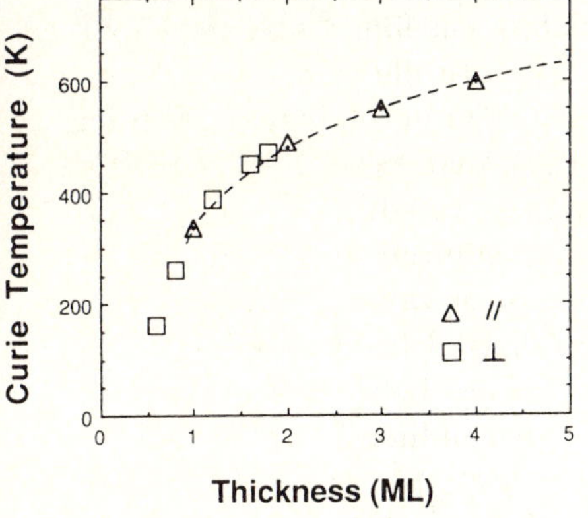

Figure III.4. Normalized $M-vs.-T$ curves for Fe/Pd(100). The top panel is for in–plane easy axes (300 K growth), and the bottom panel is for vertical easy axes (100 K growth).

Figure III.5. Log–log plots for Fe/Pd(100). The fitted β–values compare favorably with the two–dimensional Ising value of $\beta_C=1/8$.

Figure III.6. T_C vs. thickness for Fe/Pd(100). The dashed line connects the two–dimensional Ising–model results of Binder and Hohenberg for integral layer thicknesses, normalized to experiment at 3 ML.

Ising models. Their results for integral layer thicknesses, normalized to experiment at 3 ML, are shown connected by the dashed curve. There is quite good agreement between experiment and calculation. Recent theoretical work by Bander and Mills [III.14] predicted that T_C would decrease in the vicinity of the critical thickness value for vertical easy axes, based on a two–dimensional

Heisenberg model with spin–orbit–only anisotropy. The monotonic increase of the experimental T_C–values, however, is contrary to their prediction and suggests that multiple anisotropies are at play.

III.5. Summary

Three issues in surface magnetism are addressed using SMOKE characterizations of the magnetic properties of the epitaxial systems Fe on Cu(100), Ru(0001), and Pd(100). The first issue concerns the criteria for impediments to achieving monolayer magnetism. The second involves the nature and origin of the surface magnetic anisotropy. Both of these issues involve the competing influence of electronic and geometric structural dimensionality more straightforwardly. Taken together the three illustrate the richness of opportunities and challenges in the field of low–dimensional magnetism.

This work was supported by the U. S. Department of Energy, Basic Energy Sciences–Material Sciences, under contract W–31–109–ENG–38.

References

III.1 E. R. Moog and S. D. Bader, Superlattices Microstruct. **1**, 543 (1985); S. D. Bader, E. R. Moog, and Grünberg, in *Magnetic Properties of Low Dimensional Systems*, Springer Proceedings in Physics 14, edited by L. M. Falicov and J. L. Morán–López (Springer–Verlag, Berlin, (1986) p. 70.

III.2 C. Liu, E. R. Moog, and S. D. Bader, Phys. Rev. Lett. **60**, 1875 (1988).

III.3 T. Beier, H. Jahrreiss, D. Pescia, Th. Woike, and W. Gudat, Phys. Rev. Lett. **61**, 1875 (1988).

III.4 J. Araya–Pochet, C. A. Ballentine, and J. L. Erskine, Phys. Rev. B **38**, 7846 (1988).

III.5 S. H. Lu, J. Quinn, D. Tian, F. Jona, and P. M. Marcus, preprint.

III.6 D. Tian, S. C. Wu, F. Jona, and P. M. Marcus, Solid. State Commun. (1989) in press.

III.7 M. Stampanoni, A. Vaterlaus, M. Aeschlimann, F. Meier, and D. Pescia, J. Appl. Phys. **64**, 5321 (1988).

III.8 S. D. Bader and E. R. Moog, J. Appl. Phys. **61**, 3729 (1987).

III.9 W. F. Egelhoff, Jr., and I. Jacob, Phys. Rev. Lett. **62**, 921 (1989).

III.10 W. A. A. Macedo and W. Keune, Phys. Rev. Lett., **61**, 475 (1988).

III.11 A. M. Clogston, B. T. Matthias, M. Peter, H. J. Williams, E. Corenzwit, and R. C. Sherwood, Phys. Rev. **125**, 541 (1962).

III.12 J. Crangle, Philos. Mag. **5**, 335 (1960).

III.13 W. Dürr, M. Taborelli, O. Paul, R. Germar, W. Gudat, D. Pescia, and M. Landolt, Phys. Rev. Lett. **62**, 206 (1989).

III.14 M. Bander and D. L. Mills, Phys. Rev. B **38**,12015 (1988).

III.15 K. Binder and P. C. Hohenberg, IEEE Trans. Magn., MAG–**12**, 66 (1976).

IV. Dead Layers in Thin-Film Magnetism: p(1 × 1)Ni on Ag(100) and Ag(111)

J. Araya-Pochet[1], C.A. Ballentine[2], and J.L. Erskine[2]

[1]Escula de Fisica, Universidad de Costa Rica, San José, Costa Rica
[2]Department of Physics, University of Texas, Austin, TX 78712, USA

Magneto–optical techniques are used to probe layer–dependent magnetic properties of ultra–thin Ni films on Ag(100) and Ag(111) substrates. A p(1×1) Ni monolayer (n=1) on Ag(111) is non-magnetic at 30 K whereas a monolayer Ni film on Ag(100) is ferromagnetic at 110K. The preferred spin orientation of thin (2–5 layer) Ni films is in the film plane on both Ag(111) and Ag(100). The striking onset of magnetism at $n > 1$ for p(1×1)Ni on Ag(11) suggests *sp–d* hybridization plays a dominant role in ultra–thin film magnetism of transition metal thin–film systems.

IV.1. Introduction

One of the more interesting results of early attempts to explore thin–film magnetism was the reported observation of magnetically "dead" layers by Lieberman *et al.* [IV.1]. The experiments were conducted on Ni films electroplated on copper rather than using the clean ultra–high vacuum epitaxy techniques generally used in more recent studies, and therefore the possibility that the quenching of magnetism in the thin films by various effects related to surface oxidation or hydrogenation cannot be ruled out. However, these pioneering studies clearly stimulated a great deal of subsequent experimental and theoretical investigations of surface– and thin–film magnetism.

More recent experiments on carefully prepared and characterized thin magnetic layers have yielded results which, in some cases, contradict the work of Lieberman, but in other cases still suggest that magnetic dead layers can occur in thin film systems. Electron capture [IV.2] and spin–polarized photoemission [IV.3] experiments of thin Ni films on Cu substrates failed to detect any evidence of a dead layer of thin (two–layer) films. The interpretation of these experiments did involve certain assumptions required to extrapolate results to single–layer films. Other experiments by Bergmann [IV.4] and Meservey [IV.5,IV.6] did suggest quenching of magnetic moments at interfaces in several cases. Specifically, Ni films were shown to exhibit a transition from ferromagnetism to paramagnetism as the film thickness (on both Al [IV.4] and Pb–Bi [IV.5] substrates) was reduced below 2–3 atomic layers. In corresponding experiments for Fe layers [IV.4,IV.5], the magnetic moment persisted to submonolayers on substrates where Ni lost its moment at 2–3 atomic layers.

Springer Proceedings in Physics, Vol. 50   **Magnetic Properties of Low-Dimensional Systems II**

Editors: L.M. Falicov · F. Mejía-Lira · J.L. Morán-López © Springer-Verlag Berlin, Heidelberg 1990

These results appear to have stimulated several systematic theoretical studies, based on first–principles calculations, of the role a (nonmagnetic) substrate plays in governing magnetic behavior of ultra–thin epitaxial films. In these systems several factors can play an important role in determining magnetic phenomena including reduced coordination, two–dimensional symmetry, and the electronic properties that are intrinsic to surfaces and interfaces, i.e. surface and interface electronic states. Calculations by Wang and Freeman [IV.7], and by others [IV.8–IV.12] for Ni monolayer films on Cu(100) all agree that the film should be ferromagnetic. Fu and Freeman [IV.11] have shown that noble–metal substrates appear to play an important role in promoting enhanced two–dimensional magnetism in epitaxial transition–metal films and superlattice structures. Tersoff and Falicov [IV.9], in an attempt to explain disparate experimental results, carried out an extensive investigation that emphasized the role of the substrate on the magnetic properties of Ni films of 1 to 5 atomic layers on Cu(100) and Cu(111). One specific result of their study that has particular relevance to results reported in this paper is the prediction that a Ni monolayer is magnetic on Cu(100) but paramagnetic on Cu(111).

We present a study of the layer–dependent magnetic properties of Ni thin films on Ag(100) and Ag(111) surfaces. Our results verify that Ni films ranging in thickness from 1–5 atomic layers on Ag(100) are ferromagnetic with spin–moment in the plane of the film. Corresponding films on Ag(111) exhibit the same preferred spin orientation, except that the magnetism vanishes at one atomic layer. This result demonstrates the existence of a magnetic "dead" layer, as predicted by Tersoff and Falicov [IV.9], in a case where the only apparent difference between a magnetic and nonmagnetic layer is the coordination (four–fold rather than six–fold) and subtle differences in sp–d hybridization associated with different crystal surfaces.

IV.2. Experiment

Our epitaxial films were prepared and studied *in situ* under ultra–high vacuum conditions (below 1×10^{-10} torr). Conventional techniques were used to prepare the 1 cm diameter Ag(100) and Ag(111) crystals, and to clean the surfaces prior to growing the epitaxial layers (repeated Ne$^+$ sputtering at 500 eV, 1 μA followed by annealing). Thin films were grown by electron–beam evaporation (at a rate of $\sim$0.1 sec) from the tip of a 99.95% pure 0.08" diameter Ni wire. The crystal was held at temperatures ranging from 110 K to 300 K during sample preparation to suppress (and study) possible interdiffusion at the interface. Crystal order and chemical purity of the Ag substrates and the epitaxial films were monitored by low–energy electron diffraction and Auger spectroscopy.

Because of the nature of our primary result (first layer of Ni on Ag(111) has been determined to be magnetically dead), before continuing our discussion, we briefly mention several important matters related to our thin–film preparation and film structures. Surface contamination by O, C, and S impurities of the substrate surfaces, and of the grown films was carefully monitored by Auger analysis, and found to be below 1%. Film–thickness values indicated in our figures are based on absolute calibrations involving a pair of quartz microbalances. The thickness calibrations also agree with breaks in the slope of Auger peak intensities plotted as a function of time during film growth which indicate the completion of successive layers.

Good p(1×1) LEED patterns persist as Ni atoms are deposited onto Ag(111) up to about seven layers. This result (along with our Auger results) is consistent with previous studies [IV.13] of Ni epitaxial growth on Ag(111) which have shown that Ni exhibits pseudomorphic layer growth in ideal registry with the substrate lattice. Analysis of time dependent Auger signals during film growth suggest that Ni grows in a "terrace" mode on Ag(111) in which second layer atoms form before the first layer is complete. Our films were grown on cooled surfaces (110 K) to prevent interdiffusion. However, it appears that a monolayer of Ni is stable on Ag(111) at temperatures above 500 K. The epitaxy of Ni on Ag(100) is less favorable. At 110 K, a monolayer of Ni atoms destroys the LEED pattern. Previous TEM studies [IV.14] of Ni growth on Ag(100) at low temperature (93 K) have shown that the large misfit between Ni and Ag is accommodated through the lattice expansion of Ni and stacking faults up to 150 Å thickness where pseudomorphic growth of Ni(100) is established. Our timed Auger studies do indicate that significant clustering does not occur at 1 ML. Breaks in time dependent Auger intensity correlate well with our film–monitor–based first– and second–layer calibrations, suggesting that Ni films nucleate rather than clusters. Fortunately, the interesting effect occurs for Ni on Ag(111), which is well behaved.

We use the surface magneto–optic Kerr effect (SMOKE) to probe the magnetic properties of the thin films. We have used this technique to investigate the thickness– and temperature–dependent spin–anisotropy of epitaxial films of Fe on Ag(100) [IV.15]. We note that SMOKE signals (the rotation and ellipticities) produced by a monolayer of Ni atoms are significantly smaller than corresponding signals from Fe films.

Fig. IV.1 displays our magneto–optic Kerr effect data for Ni layers on Ag(100) obtained with H applied in the film plane. The films are clearly ferromagnetic starting at one ML and exhibit a large coercive force. No Kerr effects signals were detected for 1 ML films when H was applied normal to the film. This result proves that the spin–orbit anisotropy favors in–plane spin orientation at 1 ML as predicted by Gay and Richter [IV.16]. Films greater than 10 ML, begin to exhibit polar Kerr effects. This behavior can be attributted to a film–morphology–induced shape anisotropy. TEM studies of Ni films deposited (near room temperature) have shown that films less than 50 Å thick form truncated pyramids or flat–topped islands which can induce a large shape anisotropy perpendicular to the film plane.

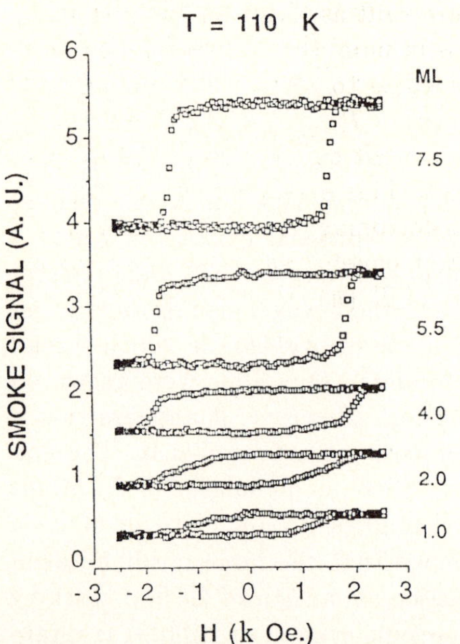

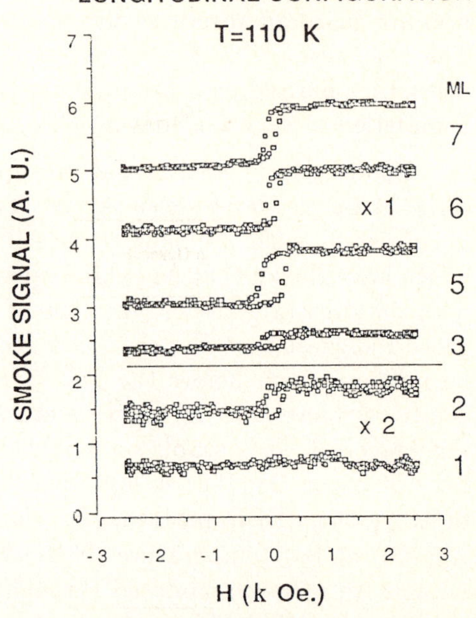

Figure IV.1. SMOKE derived hysteresis loops for ultrathin Ni films deposited on Ag(100) surfaces. Film thickness is indicated at right of each loop in monolayers (ML). The applied field is parallel to the film surface (longitudinal configuration).

Figure IV.2. SMOKE derived hysteresis loops for ultrathin Ni films deposited on Ag(111) surfaces. Film thickness is indicated at the right of each loop in monolayers (ML). The applied field is parallel to the film surface (longitudinal configuration).

Fig. IV.2 displays longitudinal configuration Kerr effect results for epitaxial p(1×1) Ni films on Ag(111). We note that the rotation for a bilayer of Ni atoms on Ag(111) is approximately 5 seconds of arc at 633 nm, which is about 1/10 that for Fe on Ag(100). The Kerr effect amplitude for a 1 ML film is at least 10 times smaller. The coercive force for Ni films on Ag(111) that exhibit ferromagnetic behavior (greater than two layers) is small (200 Oe) just as observed for other monolayer epitaxial films [IV.15]. To examine the possibility that spins prefer perpendicular alignment at 1 ML on Ag(111) (as in the case for Fe on Ag(100)) we studied the polar SMOKE signal *vs*. Ni thickness using applied fields up to 2.8 kOe. No signals were detected at 1 ML. The polar SMOKE signal presents a linear dependence with the applied filed which is consistent with the shape of an unsaturated magnetization with the applied field in a hard direction.

Dead layers can originate on the basis of a thickness–dependent transition temperature $T_c(n)$. In order to test this possibility, we study the thickness

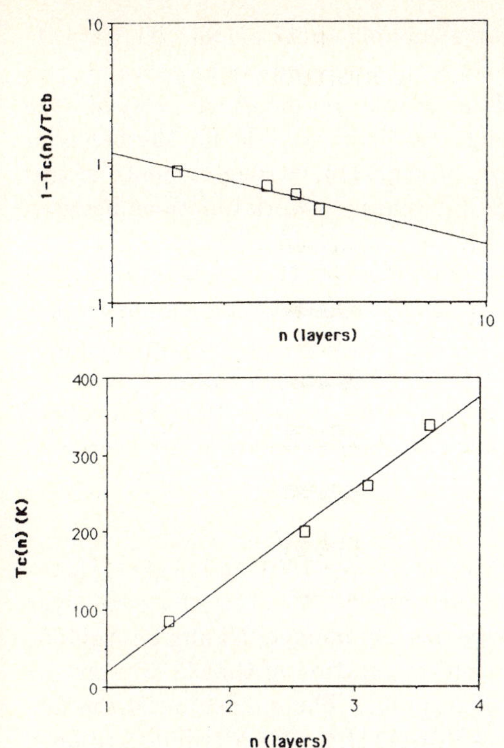

Figure IV.3. Upper panel shows the normalized transition temperature shift as a function of thickness in number n of layers, which according to Allan [IV.17] follows $[T_C(n) - T_{Cb}]/T_{Cb} = Cn^{-\lambda}$ where λ is the shift exponent and C is a constant. This extrapolation predicts a non-ferromagnetic behavior for $n = 1$. Lower panel shows a linear extrapolation of $T_C(n)$ as a function of n.

dependence of T_C and find, as seen in Fig. IV.3, that a linear extrapolation yields a non-zero value for $T_C(n = 1)$. However a more accurate extrapolation based upon the Allan's [IV.17] calculation for the transition temperature shifts $T_C(n) - T_{Cb}$, where $T_{Cb} = T_C(\infty)$, yields a negative value for $T_C(n = 1)$ indicative of a non–magnetic behavior.

IV.3. Discussion

The most striking features of these experimental results are the absence of ferromagnetism for 1 ML of Ni on Ag(111) and the sudden onset of magnetism at 2 ML. Notwithstanding the lack of good epitaxy on the Ag(100) surface, the primary differences between 1 ML of Ni on Ag(100) and on Ag(111) are the two–dimensional coordination between Ni atoms (four fold for p(1×1) Ni/Ag(100) and six fold for p(1×1) Ni/Ag(111) and subtle electronic effects associated with the different crystal faces. The Ni–Ni separation is the same on both substrates. The calculations of Tersoff and Falicov [IV.9] show that the most significant factor acting to suppress magnetization of an epitaxial layer of Ni on Cu(111) is sp–d hybridization which changes the shape of the band edge. This effect is particularly important in Ni because the Ni d–band is nearly full, and the Fermi level is close to the upper band edge. Small changes at the band edge can significantly reduce the effective number of d holes, and quench the magnetic moment. Since we have found that a monolayer of p(1×1) Ni/Cu(111) is ferromagnetic [IV.18], Tersoff and Falicov's specific predictions pertaining to the Ni film magnetic moment on Cu(111) appear to be incorrect, however, the general concept of sp–d induced quenching could still account for the behavior that we observed for p(1×1) Ni monolayer on Ag(111). Strong evidence of this mechanism is provided by considering the differences in work functions between Ni(111) and Ag(111) (0.61 eV) and Ni(100) and Ag(100) (0.58 eV) [IV.19]; a larger value indicating a more significant amount of charge transfered to the Ni overlayer. This effect would explain the fact that the transition temperature of the monolayer, which is proportional to the coordination number, is lower, in the Ni/Ag(111) system than in the Ni/Ag(100) system. The lowering of the transition temperature can be explained in the present case by the weakening of the exchange interaction due to charge transfer between substrate and overlayer.

IV.4. Summary

The magnetic properties of ultrathin Ni films on Ag(100) and Ag(111) substrates have been studied by SMOKE magnetometry. A p(1×1) Ni monolayer on Ag(111) is non ferromagnetic at 30 K whereas a monolayer Ni film on Ag(100) is ferromagnetic at 110 K. The preferred spin orientation of thin (2–5 layer) Ni films is in the plane on both Ag(111) and Ag(100). The quenching of the ferromagnetic behavior in the monolayer of Ni/Ag(111) can be attributed to sp–d hybridization.

The authors wish to thank Richard Fink for assistance in conducting these experiments. This work was sponsored by the Joint Services Electronic Program AFOSR–F–49620–86–C–0045, and National Science Foundation DMR–87–02848. One of the authors (JAP) wishes to acknowledge support from CONICIT.

References

IV.1 L. N. Liebermann, J. Clinton, P.M. Edwards, J. Mathon, Phys. Rev. Lett. **25**, 232 (1970).

IV.2 C. Rau, Comments Solid State Phys., 9,177 (1980).

IV.3 D. T. Pierce and H. C. Siegmann, Phys. Rev. B **9**, 4035 (1974).

IV.4 G. Bergman, Phys. Rev. Lett. **41**, 264 (1978).

IV.5 R. Meservey, P. M. Tedrow, V. R. Kalvey, Solid State Commun. **30**, 969 (1980).

IV.6 J. S. Moodera and R. Meservey, Phys. Rev. B **29**, 2943 (1984).

IV.7 C. S. Wang and A. J. Freeman, Phys. Rev. B **21**, 4585 (1980).

IV.8 D. S. Wang, A. J. Freeman, and H. Krakauer, Phys. Rev. B **24**, 1126 (1981).

IV.9 J. Tersoff and L. M. Falicov, Phys. Rev. B **26**, 6186 (1982).

IV.10 H. Huang, X–Y, Zhu, and J. Hermanson, Phys. Rev. B**29**, 2270 (1984).

IV.11 C. L. Fu, A. J. Freeman, and T. Oguchi, Phys. Rev. Lett. **54**, 2700 (1985).

IV.12 S. Blügel, M. Weinert, and P. H. Dederichs, Phys. Rev. Lett. **68**, 1077 (1988).

IV.13 M. G. Barthes and A. Rolland, Thin Solid Films **76**, 45 (1981).

IV.14 L. A. Bruce and H. Jaeger, Philos. Mag. **36**, 1331 (1977).

IV.15 J. Araya Pochet, C. A. Ballentine, and J. L. Erskine, Phys. Rev. B **38**, 7846 (1988).

IV.16 J. G. Gay and R. Richter, Phys. Rev. Lett. **56**, 2728 (1988); J. Appl. Phys. **61**, 3362 (1987).

IV.17 A. Allan, Phys. Rev. B **1**, 352 (1970).

IV.18 C. A. Ballentine, R. L. Fink, J. Araya Pochet, and J. L. Erskine. To be published.

IV.19 H. B. Michaelson, J. Appl. Phys. **48**, 4731 (1977).

V. Study of Short-Range Magnetic Order Transitions by Spin-Polarized Photoelectron Diffraction

C.S. Fadley [V.1]

Department of Chemistry, University of Hawaii, Honolulu, HI 96822, USA

The use of spin polarized photoelectron diffraction (SPPD) for studying short–range magnetic order is introduced. Atomic–like core–level multiplets are used as internal sources of spin–polarized electrons. SPPD effects have so far been observed for the two antiferromagnets $KMnF_3$ and MnO, for which it is found that short–range order exhibits an abrupt transition at temperatures of approximately 3–5 times the Néel temperature. The measurement and interpretation of such SPPD effects are discussed.

V.1. Introduction

Photoelectron diffraction (PD) and its close relative Auger electron diffraction (AED) are core–level–based techniques which can provide two important types of information in surface magnetism [V.2,V.3]:

• the atomic structure and degree of strain or disorder within the first layers of a surface or an epitaxial film,

• the type of magnetic short–range order present.

In standard PD or AED, the largely Coulomb scattering of the outgoing electrons leads to diffraction features and, for higher energies, also to simply interpretable forward–scattering peaks that can be directly used to determine structure, strain, and disorder. This capability has been illustrated in recent examples of epitaxial NiO growth on Ni(001) with and without strain [V.2,V.4], and of a temperature–dependent surface disorder transition on Ge(111) [V.5]. These and other studies have also shown that PD and AED are short–range probes primarily sensitive to the first few spheres of neighbors around a given emitter.

In what has been termed spin–polarized photoelectron diffraction (SPPD), the fundamental idea is to use core–level multiplet splittings to produce *internally-referenced* spin–polarized sources of photoelectrons that can subsequently scatter from arrays of ordered magnetic moments in magnetic materials. Fig. V.1(a) illustrates such a splitting for high–spin Mn^{2+} and indicates its intraatomic origin in simple *LS* terms. The net effect is to cause the peaks in this doublet to be very highly spin polarized, with one spin up (^5S) and the other spin down (^7S) as measured *relative to the spin of the emitting atom*. The large exchange interaction between the highly overlapping *3s* and *3d* electrons is responsible for the easily resolvable splitting of 6.7 eV between the ^5S and ^7S

Springer Proceedings in Physics, Vol. 50   **Magnetic Properties of Low-Dimensional Systems II**
Editors: L.M. Falicov · F. Mejía-Lira · J.L. Morán-López © Springer-Verlag Berlin, Heidelberg 1990

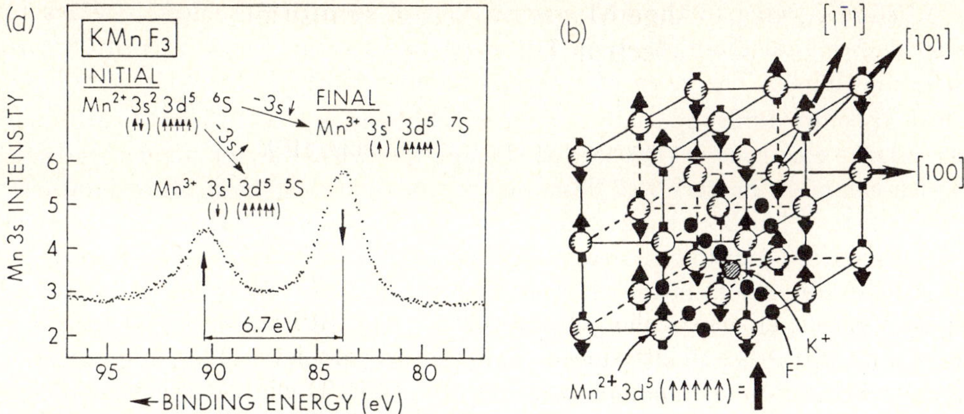

Figure V.1. (a) The multiplet–split Mn *3s* spectrum of $KMnF_3$, with the initial and final LS states and the predominant spin orientations indicated. (b) The perovskite structure of $KMnF_3$, with the antiferromagnetic ordering of the Mn^{2+} ions occurring below T_N also indicated.

final states of this photoemission process. In this free–ion picture, the resulting photoelectron peaks are furthermore predicted to be highly polarized: 100% up for ^5S and 71% down for ^7S [V.6].

In SPPD, one thus looks for spin–dependent scattering effects that make two such peaks behave slightly differently in the presence of a magnetically ordered set of scatterers. Such effects were first discussed by Sinkovic and Fadley [V.6a] from a theoretical point of view, and it was predicted that measurable effects of up to 5–10% should be seen. It is important to note that there is no need for any kind of external spin detector beyond that achieved via an electron spectrometer capable of resolving the two peaks in energy. Also, the fact that the photoelectron spins are referenced to that of the emitting atom/ion means that SPPD should be capable of sensing magnetically ordered scatterers even when the specimen has no net magnetization. Thus, studies of both ferromagnetic and antiferromagnetic materials should be possible, and meaningful measurements should also be feasible above the relevant macroscopic transition temperatures (Curie or Néel temperatures, respectively). The photoelectron emission process is also very fast, with a time scale of only about 10^{-16} to 10^{-17} seconds; thus, SPPD should provide an effectively instantaneous picture of the spin configuration around each emitter, with no averaging due to spin–flip processes. Finally, the previously discussed strong sensitivity of any form of photoelectron diffraction to the first few spheres of neighboring atoms means that SPPD should be a probe of *short–range magnetic order* (SRO). Thus, provided that a sufficiently well characterized and resolved multiplet exists for a given material, SPPD has considerable potential as a rather unique probe of SRO for a broad variety of materials, surface structures, and temperatures.

Before turning to the first observations of such SPPD effects, we ask to what degree final–state effects such as core–hole screening may alter or obscure these multiplets. The cases of principal interest in SPPD are *outer* core holes which are more diffuse spatially and for which the interaction with the surrounding valence electrons is thus not as strongly polarizing as for inner core holes (which can often be very well described in the equivalent–core approximation). Nonetheless, it has been suggested by Veal and Paulikas [V.7] that both screened and unscreened multiplets corresponding to $3d^{n+1}$ and $3d^n$ configurations, respectively, are present in the *3s* spectra of even highly ionic compounds such as MnF_2. Since such effects would make the carrying out of SPPD measurements more difficult (although still certainly not impossible) due to the potential overlap of peaks of different spin polarization, Hermsmeier *et al.* [V.8] have explored this problem in a study of Mn *3s* and *3p* multiplets for which the experimental spectra from several reasonably ionic solid compounds have been directly compared to the analogous spectra from *gaseous* Mn, a simple free–atom system in which no extraatomic screening can occur.

In Fig. V.2, we show their compilation of *3s* spectra for: (a) single–crystal MnO with (001) orientation, (b) polycrystalline MnF_2 as obtained some time ago by Kowalczyk *et al.* [V.9], (c) gaseous atomic Mn, and (d) a prior theoretical calculation of these multiplets by Bagus *et al.* which totally neglects extra

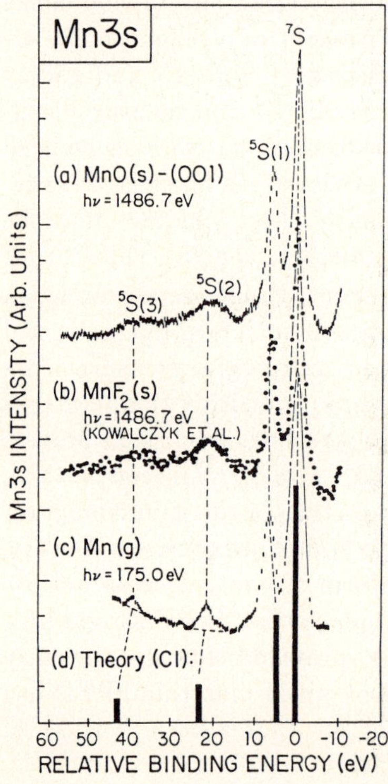

Figure V.2. Experimental Mn *3s* spectra for (a) MnO(001), (b) polycrystalline MnF_2 (from Ref. V.9), and (c) gas–phase atomic Mn are compared to (d) theoretical calculations for emission from a free Mn^{2+} ion including CI effects in the final state (from ref. V.10).

atomic screening [V.10]. From a consideration of the experimental data only, it is striking that the solid–state spectra are very similar to the gas–phase spectra, with the only differences being some extra broadening in the solid and some small changes in peak positions that are not at all surprising. This is true also for the more complex $3p$ spectra obtained in the same study [V.8]. Thus, even without resorting to theory, it seems clear that these spectra are very free–atom/free–ion like, and that a simple multiplet interpretation such as that in Fig. V.1a should rather accurately describe the spin polarizations of the photoelectrons involved.

If we consider now the best available free–ion theoretical prediction for the $3s$ spectra, this conclusion becomes even more convincing. In Fig. V.2d, the results of calculations by Bagus *et al.* [V.10] for Mn^{3+} with a $3s$ hole and limited configuration interaction (CI) are shown. There is excellent agreement with experiment for not only the two dominant members of the multiplet that would be most useful in SPPD, but also for the two much weaker satellites that directly result from including CI. Similar conclusions are reached in a comparison of experiment and theory for analogous $3p$ spectra [V.8]. We thus conclude that extra atomic screening does not cause a major perturbation of these multiplet splittings. This implies that outer core holes such as $3s$ and $3p$ in $3d$ systems or $4s$, $4p$, and $4d$ in $4f$ systems should exhibit relatively free–atom/free–ion like multiplets for a variety of high–spin cases. Such multiplets should in turn be useful as spin–resolved sources in SPPD.

Turning now to the question of final–state effects in core spectra for the more complex case of a ferromagnetic metal, we note that a recent direct measurement with an external spin detector of the spin polarization over the Fe $3p$ peak from ferromagnetic Fe by Kisker and Carbone [V.11] also yields significant spin–up polarization at lower kinetic energy and spin–down polarization at higher kinetic energy. These polarizations have the same qualitative sense as those in Fig. V.2; they are also in the directions expected for a simple free–atom model of the $3p$ multiplet splitting [V.8]. These results thus suggest that SPPD should be possible with ferromagnetic metals as well.

V.2. Experimental Results for $KMnF_3$ (110)

Turning now to a consideration of the relatively few SPPD experiments carried out to date, we show in Fig. V.1b the crystal structure of the first material for which such effects were observed: a (110)–oriented crystal of the simple antiferromagnet $KMnF_3$. It is clear that the relative spins of the emitter and the first scatterer encountered can be different for different directions of emission, as for example between [100] and [101]. Spin–dependent scattering effects were first observed for this system by Sinkovic *et al.* [V.12] as small changes of up to about 15% in the ratios of the $^5S(1)$ = spin–up and ^7S = spin–down peaks in the dominant doublet. For this study, a lower energy of excitation of 192.6 eV (Mo $M\varsigma$ radiation) was used in order to yield lower energy photoelectrons of approximately 100 eV and a concomitantly greater influence of spin–dependent effects in scattering [V.6]. (SPPD is thus a type of measurement inherently suited to synchrotron radiation in the soft–X–ray range.) The $^5S(1):^7S$ intensity ratio was found to be sensitive to both *direction of emission* (as qualitatively expected from Fig. V.1(b)) and *temperature*. The latter is in fact the most striking aspect of these results, since the variation of this ratio with temperature is found to exhibit a rather sharp change at a temperature considerably above the Néel temperature (T_N), as shown in Fig. V.3a. Here, an experimental spin asymmetry S_{expt} that is measured relative to the value of the ratio at a high–temperature limit is plotted against temperature for two different directions of emission. For spin–up/spin–down ratios of R_{LT} at the temperature of interest and R_{HT} at the high–temperature limit, S_{expt} (%) is defined as $100[(R_{LT}-R_{HT})/R_{HT}]$. The high–temperature drops observed here in the spin asymmetry have been suggested to be due to some sort of abrupt reduction of the short–range magnetic order that is expected to dominate in producing such spin–polarized photoelectron diffraction effects [V.12]. Changing the emission direction by only 9° is also seen in Fig. V.3a to change the form and magnitude of the transition significantly (note that the right and left scales are different).

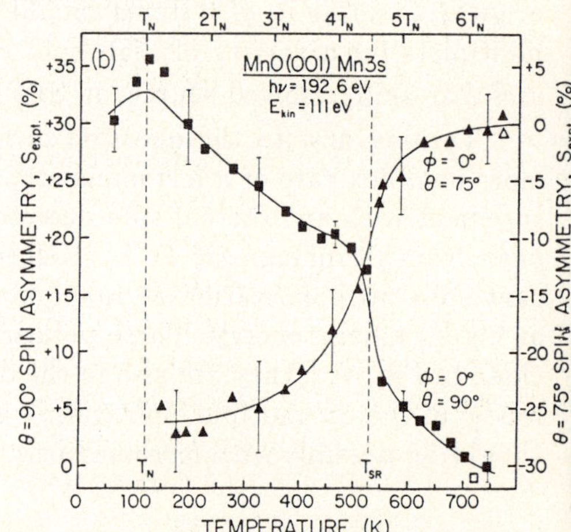

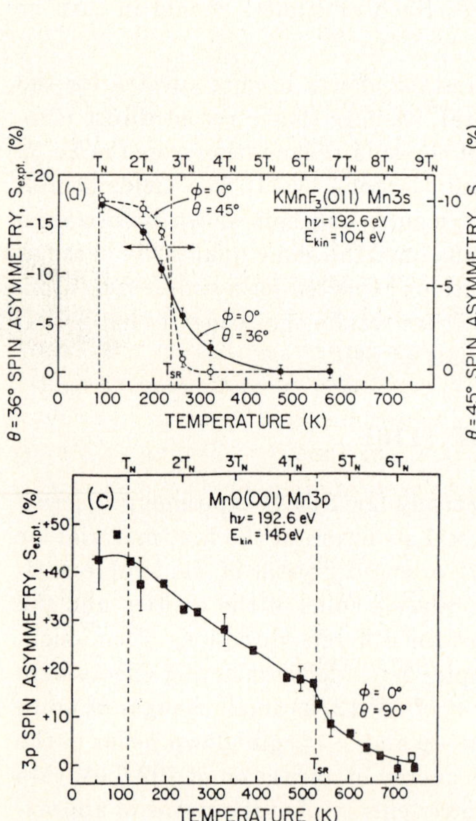

Figure V.3. The temperature dependence of experimental spin asymmetries for three cases, with electron emission along $\theta = 90°$ corresponding to normal emission. Mo Mς radiation at 192.6 eV was used for excitation. Cases are: (a) Mn *3s* emission from KMnF$_3$ (011) (from Ref. V.12), (b) Mn *3s* emission from MnO(001) (from Ref. V.13a), and (c) Mn *3p* emission from MnO(001) (from Ref. V.13b).

V.3. Experimental Results for MnO(001)

In an important confirmation and extension of this earlier work, very similar SPPD effects have also more recently been observed by Hermsmeier *et al.* for (100)–oriented MnO [V.13], and two of their curves for the temperature dependence of the Mn *3s* spin asymmetry are shown in Fig. V.3b. As for $KMnF_3$, there is a relatively sharp change in the ${}^5S(1){:}{}^7S$ ratio at a temperature that is again well above T_N . For MnO it is also interesting that the direction of this change reverses upon changing emission direction only slightly. (Note also the different directions on the axes of Figs. V.3a and V.3b).

A very similar transition at the same temperature has also been seen for MnO in the *3p* multiplet, where the more widely split ${}^5P(1){:}{}^7P$ ratio is monitored [V.13b]. This is shown in Fig. V.3c. The fact that two different multiplets exhibit the same sort of transition in the ratio in spite of being at different kinetic energies (about 108 eV average for *3s* and 135 eV average for *3p*) and having different spin–up/spin–down splittings (6.5 eV for *3s* versus 17.0 eV for *3p*) strongly indicates that this is a magnetic transition, and permits ruling out other possible causes such as some sort of purely structural transition. Another observation pointing to a magnetic effect is that the *3s* ratios of Fig. V.3b show no change whatsoever with temperature when the excitation energy is raised to Al $K\alpha$ radiation at 1486.7 eV [V.13b], as expected for magnetic scattering [V.6a].

Other photoelectron diffraction measurements on MnO(001) permit ruling out the following as causes of these transitions: changes in surface composition, surface reconstruction involving only atomic position changes, and simple Debye–Waller effects [V.13b].

It is also interesting in Figs. V.3b and V.3c that both the *3s* and *3p* spin asymmetries exhibit a weak peak at the Néel temperature which suggests some sensitivity of SPPD also to long–range order (LRO).

Recent synchrotron radiation measurements on the same system by Tran *et al.* [V.14] confirm the transitions seen in Fig. V.3b, and also show that, for a fixed direction of emission and variable photon energy, the sign of the step occurring at the transition temperature is a strong function of energy. This is expected, since varying energy will vary the phases of different scattered waves, much as in an EXAFS experiment.

A further interesting observation is that the short–range–order transition temperatures (TSR) are for both $KMnF_3$ and MnO approximately equal to the Curie–Weiss temperatures of the two materials, a connection which may be associated with the fact that this constant is proportional in mean–field theory to the sum of the short–range magnetic interactions [V.15].

V.4. The Theory of SPPD

A number of interesting questions are raised by these results. We comment briefly below on two fundamental aspects of the theoretical interpretation of them.

V.4.1 Nature of the Short–Range Order Transition

The relatively sharp breaks in the spin asymmetry curves of Fig. V.3 are not consistent with any known theory of the behavior of bulk short–range magnetic order at temperatures well above T_N. For example, the typical behavior of spin–spin correlation functions in mean–field models is an exponential–like decay as temperature increases, with the only abrupt changes occurring at the LRO transition temperature [V.15].

The SPPD measurement is very surface sensitive, probing primarily the first few layers of atoms, and this could mean that the transition observed is unique to the surface and displaced significantly to higher temperatures. However, the sharpness of the transition for some cases (e.g. $\theta = 45°$ in Fig. V.3a) and its relatively large magnitude of up to 15% are not consistent with the expected averaging of different transition temperatures over different layers. Also, the fact that only a very small effect is seen at T_N implies that LRO effects are only weakly seen in SPPD, as expected because of its SRO sensitivity. So only a very high temperature SRO transition limited to one surface layer would explain these results, which seems unlikely. Thus, we are left with the most likely current hypothesis that there is a more bulk–like SRO transition which occurs at very nearly the same temperature over the first 4–5 layers of the sample, and that local anti–ferromagnetic order is strongly maintained in the first 2–3 spheres of neighbors of any emitter up to this temperature.

One characteristic possessed by both $KMnF_3$ and MnO is that both exhibit frustration in the anti–ferromagnetic ground state due to the presence of ferromagnetic interactions to both nearest neighbors (NN) and next–nearest neighbors (NNN). It is thus possible that this frustration is responsible for the abrupt change seen in SRO at high temperatures.

As one indication of what a mean–field approach says about such a frustrated system, we show in Fig. V.4 theoretical curves for different spin–spin correlation functions σ as calculated by Sánchez and Morán–López [V.16]. Here, $\sigma_1 = $ first neighbor, $\sigma_2 = $ second neighbor, and $\sigma_3 = $ third neighbor; the system is $KMnF_3$ with $J_2 = 0.2\ J_1$. As expected, these curves show the usual exponential–like decay above T_N and no abrupt changes at high temperature. However, it is interesting in Fig. V.4b that both σ_2 and σ_3 show sign changes with increasing temperature, with σ_2 crossing zero at a temperature of about 330 K that is of the same order as the experimental TSR of 240 K. In addition, σ_2 and σ_3 cross one another exactly at 240 K. Moreover, in preliminary calculations for this system including more neighbors in the mean–field statis-

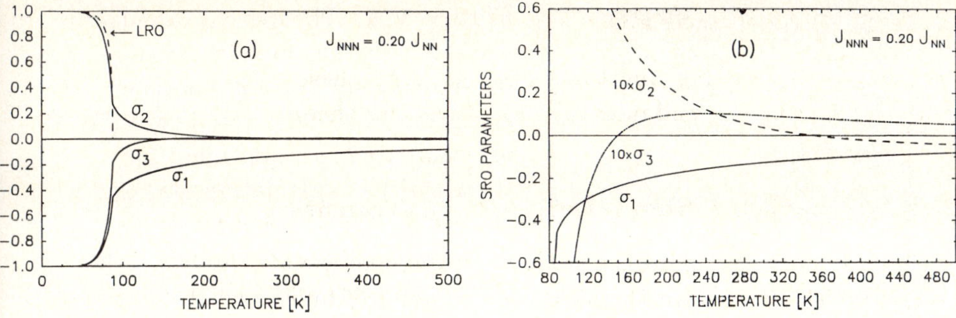

Figure V.4. Spin correlation functions calculated by Sánchez and Morán–López (Ref. V.16) for $KMnF_3$ in the presence of frustration and with $J_2 (J_{NNN})=0.2\, J_1 (J_{NN})$. Shown are σ_1, σ_2, and σ_3 for first, second, and third neighbors, respectively. The LRO transition at $T_N = 88$ K is evident in (a); the SRO parameters are magnified for higher temperatures in (b).

tics, Noguera and Finel [V.17] have in fact seen an abrupt sign change in σ_2 at comparable temperatures.

Thus, although all of these calculated correlation functions are too small at high temperature to explain the magnitude of the SPPD effects observed, these results suggest that frustration may play an important role in the transitions observed. More accurate modeling would be very interesting, for example, with large–cluster Monte Carlo methods and/or a more accurate Hamiltonian for treating SRO within the first three shells of neighbors.

V.4.2 Spin–Polarized Diffraction Calculations

Diffraction calculations aimed at understanding such SPPD effects have been carried out by Sinkovic *et al.* [V.6] and by Hermsmeier *et al.* [V.13b]. In the absence of detailed knowledge of the spin correlation functions over the transition region, it is assumed that, from well below to well above T_{SR}, the system goes from local ground–state anti–ferromagnetic order to the paramagnetic state. The only magnetic scattering effect that is included is that of exchange, which is included in both the Kohn–Sham (KS) and weaker energy–dependent Dirac–Hara (DH) forms. Such calculations have been found to qualitatively predict the magnitudes and, in some cases, also the systematics, of the experimental data, as illustrated for MnO in Fig. V.5 [V.13b]. Here, experiment and theory are compared for the two types of magnetic domains expected at the MnO surface, as well as for the sum of the two appropriate to their presence with equal statistical weighting.

These results indicate that the magnitude of the SPPD transition is correctly predicted by this theory if the stronger KS exchange is used; the DH is about 3 times too weak, however. The systematics of the limited 4–point data

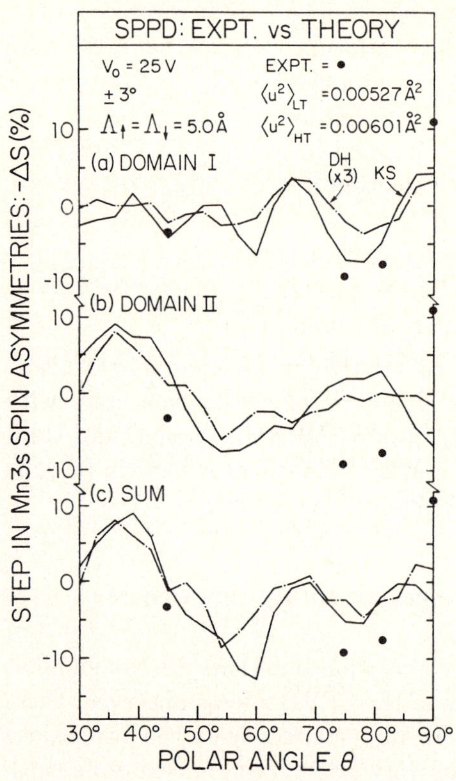

Figure V.5. Theoretical calculations of the steps in spin asymmetry on crossing T_{SR} for Mn $3s$ emission from MnO(001) (cf. Fig. V.3b) are compared to experiment for three types of calculation: (a) and (b) are for the two symmetry–inequivalent magnetic domains that can be present at the surface, and (c) is for the sum of these two with equal statistical weight (from Ref. V.13b). Some of the parameters used in the calculations are indicated at the top of the figure.

set are best predicted by the curve for Domain I alone, whereas the in–principle more correct summed curve is less satisfactory.

Based upon this and other comparisons of experiment and theory [V.6b, V.13b], it can be concluded that such spin–polarized photoelectron diffraction calculations represent a reasonable first step toward the modeling of these effects, but that comparisons to more extensive sets of experimental data are needed before any definitive conclusions can be drawn concerning the degree of their validity. Improvements in the theory to incorporate multiple scattering may also be necessary, as suggested in preliminary calculations by Kaduwela and Friedman [V.18].

V.5. Concluding Remarks

Overall, we conclude that outer core multiplets provide convenient internally referenced sources of highly polarized electrons, and that they have been used to measure SPPD effects of up to 10–15% in magnitude. A new type of SRO transition has been seen in both $KMnF_3$ and MnO. This transition may be associated with the presence of frustration in these systems, but additional experimental and theoretical work will be needed to fully understand these effects.

SPPD is thus a new type of photoelectron diffraction, but it has the potential of providing a rather unique kind of information on the short–range spin–order and spin–correlation functions around a given type of emitter site in the near–surface region of magnetic materials. Other antiferromagnetic and also ferromagnetic materials are currently being studied in order to better establish the systematics of the short–range order transition and the range of utility of the method.

As a final comment, spin–polarized Auger electron diffraction (SPAED) also should be possible, since it is known that Auger spectra from ferromagnets exhibit strong spin polarization from one part of the manifold of peaks to another [V.19]. The more complex nature of Auger spectra in general will make the *a priori* prediction of the type of spin polarization expected a more difficult problem, but for ferromagnets with net magnetization, an external spin detector could be used to first calibrate the spectrum for polarization. Then, measurements of spin–up/spin–down ratios as functions of direction and/or temperature could be performed in the same way as for the spin–split core multiplets in SPPD.

References

V.1 Work supported by the Office of Naval Research under Contract N00014–87–K–0512.

V.2 C. S. Fadley, Physica Scripta T**17**, 39 (1987).

V.3 C. S. Fadley, in Springer Series in Sol. State Sciences, Vol. 81 (Springer–Verlag, Berlin, 1988).

V.4 R. S. Saiki, A. P. Kaduwela, J. Osterwalder, C. S. Fadley, and C.R. Brundle, to appear in Phys. Rev. B.

V.5 D. J. Friedman, T. T. Tran, Y. J. Kim, G. A. Rizzi, and C. S. Fadley, to be published.

V.6 (a) B. Sinkovic and C. S. Fadley, Phys. Rev. B **31**, 4665 (1985); (b) D. J. Friedman, B. Sinkovic, J. Osterwalder, A. P. Kaduwela, B. Hermsmeier, and C. S. Fadley, to be published.

V.7 B. W. Veal and A. P. Paulikas, Phys. Rev. Lett. bf 51, 1995 (1985).

V.8 B. D. Hermsmeier et al., Phys. Rev. Lett. **61**, 2592 (1988).

V.9 S. P. Kowalczyk, L. Ley, R.A. Pollak, F. R. McFeely, and D. A. Shirley, Phys. Rev. B **7**, 4009 (1973).

V.10 P. S. Bagus, A. J. Freeman, and F. Sasaki, Phys. Rev. Lett. **30**, 850 (1973).

V.11 E. Kisker and C. Carbone, Sol. St. Commun. **65**, 1107 (1988).

V.12 B. Sinkovic, B. D. Hermsmeier, and C. S. Fadley, Phys. Rev. Lett. **55**, 1227 (1985).

V.13 (a) B. D. Hermsmeier, J. Osterwalder, D. J. Friedman, and C. S. Fadley, Phys. Rev. Lett. **62**, 478 (1989); (b) B. D. Hermsmeier, J. Osterwalder,

D.J. Friedman, B. Sinkovic, T. T. Tran, and C. S. Fadley, to be published.

V.14 T. T. Tran, J. Osterwalder, E. L. Bullock, C. Guillot, J. Lecante, and C. S. Fadley, to be published.

V.15 J. E. Smart, *Effective Field Theories of Magnetism*, (Saunders, New York, 1966) Chap. 4.

V.16 J. M. Sánchez and J. L. Morán–López, private communication.

V.17 C. Noguera and I. Nigel, private communication.

V.18 A. P. Kaduwela and D. J. Friedman, unpublished results.

V.19 M. Taborelli, R. Allenspach, G. Boffa, and M. Landolt, Phys. Rev. Lett. 56, 2869 (1986).

VI. Spin-Polarized Photoemission on Ferromagnetic Epitaxial Thin Films

F. Meier, M. Stampanoni, A. Vaterlaus, and M. Aeschlimann

Laboratorium für Festkörperphysik, ETH Hönggerberg, CH-8093 Zürich, Switzerland

The magnetization curves of epitaxial Fe/Cu(100) films of 1 monolayer thickness have no perpendicular remanence at $T = 30$ K. Comparative experiments on $Fe_{50}Cu_{50}$ show that the magnetic properties of 1 ML Fe/Cu(100) rather correspond to those of a Fe–Cu–double layer situated on the rest of the Cu–substrate. Magnetic effects of the film–substrate–interaction become evident upon sandwiching the Fe film between the Cu– substrate and a Cu–overlayer.

The surfaces of bulk ferromagnets have been investigated for a long time [VI.1]. They show peculiar properties with respect to the temperature dependence of the magnetization, the ordering temperature, the size of the magnetic moments per atom, and the geometry of the spin arrangement. There is a big discrepancy between the body of computational work and the body of supporting experimental evidence: From elementary physical principles there is no doubt that the truncation of a bulk ferromagnet results in surface–specific features, e.g., in anomalous surface magnetic moments. So far, experimentally these have not been verified in a single case. This is due to the fact that suitable experimental tools are not available or —stated differently— that the expected features are too weak to be discriminated against the background of the bulk.

It is a natural idea, then, to cut off the bulk leaving only the surface in form of a thin film. The free–standing isolated film of a few monolayers (ML) thickness does not (yet) exist. However, during the last few years it became clear that single crystalline magnetic films of extraordinary quality and reproducibility can be grown on suitable substrates [VI.2]. It is a fortunate circumstance that the technique of growing epitaxial magnetic transition–metal film is exceedingly simple: the evaporator consists of a piece of resistively heated wire made from the material to be deposited. In all experiments described in this paper the single crystalline substrate was held at room temperature. Evidently, none of the elaborate furnaces and substrate temperature–controls used in semiconductor epitaxy are necessary for preparing metal films.

Today we know that ferromagnetic films of monolayer thickness reveal truly new magnetic features unconcealed by background effects [VI.3]. Some of them are described in the following.

It has become fashionable to interpret the mere existence of ML– ferromagnetism as a miraculous event. The underlying reason is that 2–dimensional ferromagnetism should not occur for isotropic nearest–neighbor exchange except in the Ising model [VI.4]. However, already in the pioneering work of Gradmann [VI.5] and others [VI.6] no doubt was left about the existence of monolayer ferromagnetism. This proves nothing but the fact that the exchange in these films is not isotropic. But then, a much more constructive question concerning the anisotropy is whether it leads to other noticeable effects besides allowing for long–range magnetic order.

The subject of magnetic surface anisotropy has been treated by Néel [VI.7] based on a phenomenological approach and more recently by Gay and Richter [VI.8], carrying out a direct quantum–mechanical calculation of the spin–orbit–interaction for realistic electronic structures. Interestingly, the findings are very similar: For cubic lattices the anisotropy terms enter in lower order at the surface than in the bulk: this holds for the polynomial expansion of the anisotropy energy in the case of Néel as for the perturbation expansion in the case of Ref. [VI.8]. The consequences are that 1) the surface anisotropy may exceed the bulk anisotropy by 2 orders of magnitude and 2) the surface anisotropy may be directed normal to the plane of the film. Then, for sufficiently thin films an interesting situation arises: the surface anisotropy may become capable of pulling the magnetization out of the in–plane direction —favoured by the demagnetizing energy— into the direction of the surface normal. As a consequence the easy direction in such a film is perpendicular, causing perpendicular remanent magnetization.

Appropriate tools to investigate the magnetism of thin films are various types of spin–polarized electron spectroscopies [VI.9]. The experimental results described below have been obtained with spin–polarized photoemission [VI.10]. Here, the spin–polarization P of the photoemitted electrons is measured. P is defined as $(N_\uparrow - N_\downarrow)/(N_\uparrow + N_\downarrow)$ where $N_\uparrow$ $(N_\downarrow)$ is the number of photoelectrons with spin magnetic moment parallel (antiparallel) to the surface normal. Most important, a magnetic field (up to 30 kOe) can be applied perpendicular to the film during measurements, assuring that the sample is always in a state of magnetic saturation. The temperature of the sample is variable between 30 K and 550 K. As light source the full spectrum of a Hg–Xe–lamp was used ($h\nu$=5.4 eV; photothreshold of the *fcc* Fe/Cu(100) samples: 4.7 eV).

Indirect evidence for perpendicular magnetization in epitaxial *bcc* Fe/Ag(100) films was first presented by Jonker *et al.* [VI.11]. Shortly afterwards direct evidence was found in films of *fcc* Fe/Cu(100) [VI.12] and *bcc* Fe/Ag(100) [VI.13]. Interestingly, for both films it was not the single monolayer which showed perpendicular remanence. Perpendicular remanence appeared only for films thicker than 2 ML. Fig. VI.1 and Fig. VI.2 show that the dependence of $P(H)$ on film thickness is most remarkable. In fact, it contradicts the notion that the perpendicular anisotropy —and thereby the coercivity— is a monotonically decreasing function of film thickness. Obviously, this would be the case

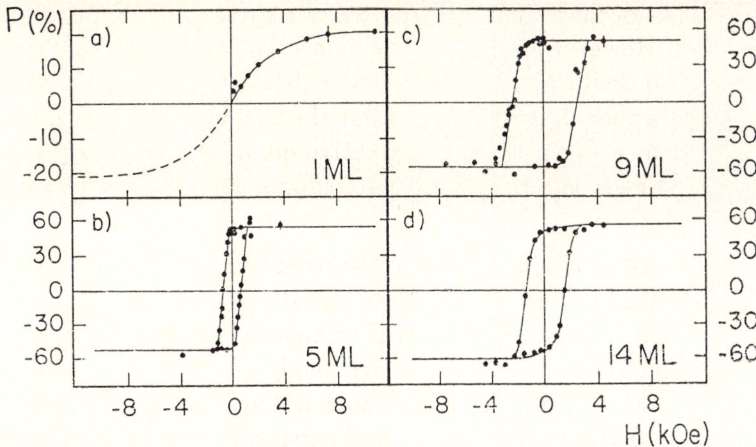

Figure **VI.1**. Threshold electron spin–polarization $P(H)$ of *fcc* Fe/Cu(100) films. The magnetic field is applied normal to the plane of the film.

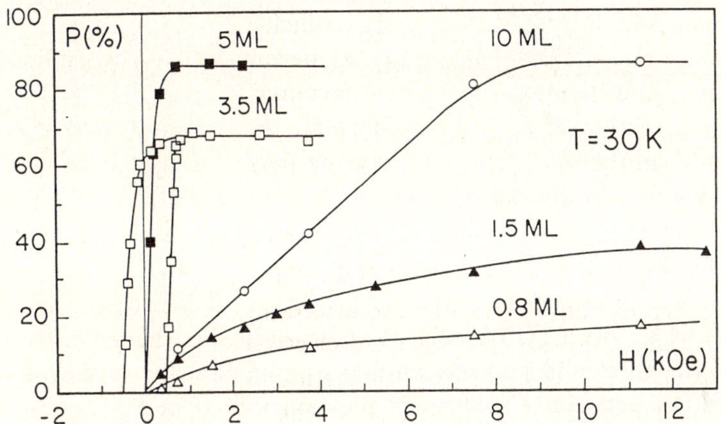

Figure **VI.2**. Threshold electron spin–polarization $P(H)$ of *bcc* Fe/Ag(100)–films. The magnetic field is applied normal to the plane of the film.

if a constant, thickness–independent surface anisotropy acted against the shape anisotropy growing proportional to the thickness of the film.

It is natural to attribute the build–up of the perpendicular anisotropy with increasing film thickness up to 2–4 ML to the film–substrate interaction. Inert substrates do not exist as has been recognized even for Ag where the d–bands are fully separated from those of the magnetic iron [VI.8]. It is tempting to consider the first ML of Fe on Cu(100) [or Ag(100)] rather as a double layer of Fe–Cu (Fe–Ag) on the rest of the substrate. Then, a straightforward and challenging extension of the experiments displayed in Fig. VI.1 is to study the magnetic effects of diluting thicker Fe films with atoms of the substrate. Ex-

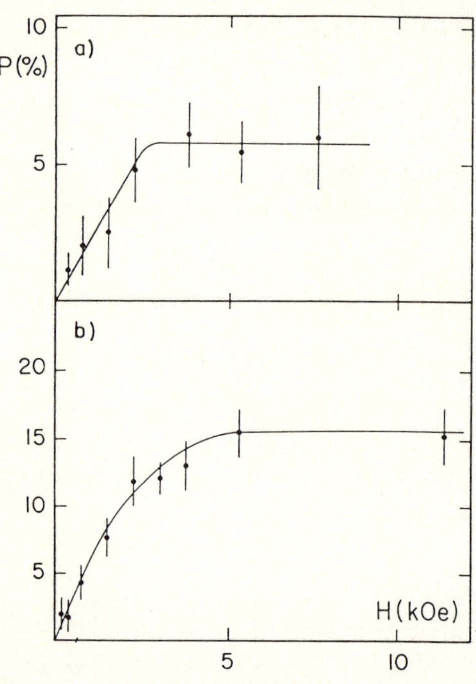

Figure VI.3. (a) Threshold electron spin–polarization $P(H)$ of 10 ML *fcc* $Fe_{50}Cu_{50}$/ Cu(100) films at $T=30$ K. The magnetic field is applied perpendicular to the film plane. (b) As (a), but for 8 ML *fcc* Fe $_{75}Cu_{25}$/ Cu(100).

perimentally, this is done by coevaporation fo Fe and Cu on Cu(100). Although the equilibrium phase diagram shows that the two elements are immiscible at room temperature, quasistable Fe_xCu_{1-x} phases do exist [VI.14].

Fig. VI.3 shows the magnetization curves of $Fe_{50}Cu_{50}$, 10 ML thick, and $Fe_{75}Cu_{25}$, 8 ML thick, at $T=30$ K. Obviously, the perpendicular remanence is no longer present and the threshold polarization has droped sharply. Although there is no direct relation between P–values at photothreshold and magnetic moments in itinerant magnets, the lower P goes together —in this case— with a smaller magnetic moment/atom in Fe_xCu_{1-x} [VI.14]. It has been found that for $Fe_{50}Cu_{50}$ and $Fe_{75}Cu_{25}$ the moments are 0.9 μ_B/atom and 1.6 μ_B/atom, respectively, resulting in magnetizations of 710 Gauss and 1260 Gauss.

For a film with only shape anisotropy the saturation field is $4\pi M_S$ where M_S is the magnetization. The value of $4\pi M_S$ for $Fe_{50}Cu_{50}$ is 8.9 kOe. This is practically identical to the saturation observed for 1 ML Fe on Cu(100), showing that in this system the perpendicular anisotropy is very weak. In contrast, the 10 ML film $Fe_{50}Cu_{50}$ saturates at 3 kOe already, thereby showing that a substantial perpendicular anisotropy is present —although not sufficient to bring about perpendicular remanence.

The $Fe_{75}Cu_{25}$ mixed film is interesting because the Fe and Cu Auger amplitudes are identical to those of a 2.6 ML Fe/Cu(100) film. However, the magnetic behaviour is entirely different: The mixed film has no remanence whereas the 2.6 ML Fe/Cu(100) film has a coercive field of 1 kOe. Obviously, this must be

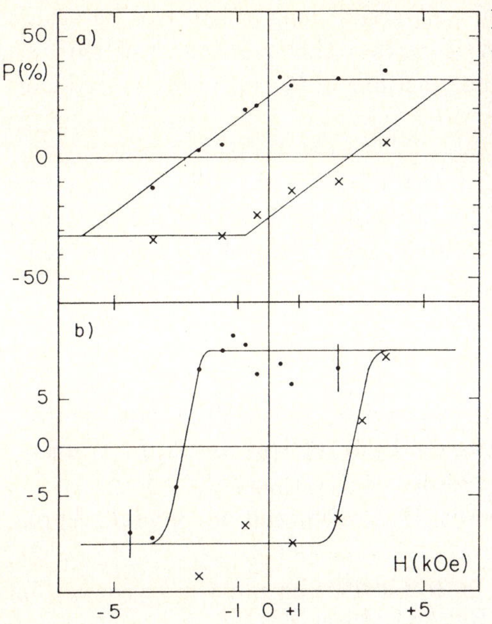

Figure VI.4. (a) Threshold electron spin–polarization $P(H)$ for a 8 ML *fcc* Fe/Cu(100) film with oblique hysteresis. $T = 30$ K. (b) As (a), but sample covered with 3 ML of epitaxially grown Cu on top of the Fe–film.

due to a marked difference in composition of these films: one is a homogeneous Fe–Cu–intermixture the other a Fe/Cu interface in *fcc* Fe/Cu(100) films [VI.15].

Direct evidence for the film–substrate interactions is given in Fig. VI.4. Sometimes, for unknown reasons, the hysteresis curves of the Fe–films are oblique as shown in Fig. VI.4a. More information is required to identify the origin of this phenomenon, in particular a direct observation of possible domain structures using spin–polarized electron microscopy [VI.16] would be helpful. Starting out from the film in Fig. VI.4a an epitaxial Cu coating (3 ML) was deposited on top of it and the influence on the form of the magnetization curve examined.

The result is shown in Fig. VI.4b: the hysteresis of the Cu/Fe/Cu sandwich has become rectangular without affecting the value of the coercive field. There is the possibility that the substrate–film interaction favors a single–domain state of the entire film —e.g., by some sort of exchange between otherwise isolated areas of Fe on terraces of the Cu–substrate— thereby inducing a magnetization curve corresponding to a single magnetic domain.

So far, the subject of thin film magnetism has yielded a great variety of strong and remarkable phenomena. The feature of perpendicular remanence is firmly established and it should be tested for use in practical applications as has been done in a preliminary study on thermomagnetic writing [VI.17]. Thin–film magnetism illustrates very impressively the interplay of magnetic, chemical and configurational properties. The newly developed local probes, such as scanning tunneling microscopy, should prove very helpful in the future [VI.18]. Then, instead of measuring spatially averaged properties, magnetism can be studied exactly at the locations of interest. It is certain that the study of thin–film model systems will help to elucidate other sections of the infinitely varied world of magnetism.

We thank H. C. Siegmann for many fruitful discussions. The expert technical assistance of K. Brunner was crucial for carryng out the experiments. The financial support by the Schweizerische Nationalfonds is gratefully acknowledged.

References

VI.1 *Magnetic Properties of Low–Dimensional Systems*, Proc. Int. Workshop, Taxco, México 6–9 Jan. 1986, ed. by L. Falicov and J. L. Morán–López (Springer–Verlag, Berlin, 1986).

VI.2 R. Germar, W. Dürr, J. W. Krewer, D. Pescia, and W. Gudat, Appl. Phys. A **47**, 393 (1988).

VI.3 M. Stamponi, Appl. Phys. A. to be published.

VI.4 C. Herring and C. Kittel, Phys. Rev. **81**, 869 (1951).

VI.5 U. Gradmann, Appl. Phys. **3**, 161 (1974).

VI.6 C. A. Neugebauer, Phys. Rev. **116**, 1441 (1962); D. Stünkel, Z. Physik **176**, 207 (1963).

VI.7 L. Néel, J. Phys. Radium **15**, 225 (1954).

VI.8 J. G. Gay and R. Richter, J. Appl. Phys. **61**, 3362 (1987); J. G. Gay and R. Richter, Phys. Rev. Lett. **56**, 2728 (1986).

VI.9 *Polarized Electrons in Surface Physics*, Adv. Series in Surface Science, ed. by R. Feder (World Scientific Publishing Co., Singapore, 1985).

VI.10 M. Campagna, D. T. Pierce, F. Meier, K. Sattler, and H. C. Siegmann, Adv. El. and El. Phys. **41**, 113 (1976).

VI.11 B. T. Jonker, K.–H. Walker, E. Kisker, G. A. Prinz, and C. Carbone, Phys. Rev. Lett **57**, 142 (1986).

VI.12 D. Pescia, M. Stamponi, G. L. Bona, A. Vaterlaus, R. F. Willis, and F. Meier, Phys. Rev. Lett. **58**, 2843 (1987).

VI.13 M. Stamponi, A. Vaterlaus, M. Aeschlimann, and F. Meier, Phys. Rev. Lett. **58**, 2843 (1987).

VI.14 C. L. Chien, S. H. Liou, D. Kofalt, Wu Yu, T. Egami, and T. R. McGuire, Phys. Rev. B **33**, 3247 (1986).

VI.15 D. A. Steigerwald and W. Egelhoff, Phys. Rev. Lett. **60**, 2558 (1988).

VI.16 D. T. Pierce, Surf. Sci. **189–190**, 710 (1988); H. P. Oepen and J. Kirschener, Phys. Rev. Lett. **62**, 819 (1989); K. Koike and K. Haykawa, Jpn. J. Appl. Phys. **23**, L187 (1984).

VI.17 M. Aeschlimann, M. Stamponi, A. Vaterlaus, and F. Meier, J. Physique C **8**, 1963 (1988).

VI.18 R. Allenspach and A. Bischof, Appl. Phys. Lett. **54**, 587 (1989).

VII. Spin-Dependent Inelastic Electron Scattering from Ferromagnetic Surfaces

H. Hopster, D.L. Abraham, and D.P. Pappas

Department of Physics and Institute for Surface and Interface Science,
University of California, Irvine, CA 92717, USA

Spin–dependent electron energy loss mechanisms in Ni are studied in detail by scattering a polarized electron beam off a Ni(110) surface and measuring the spin–polarization as a function of energy loss. For small energy losses (off specular) isotropic exchange scattering is found to be the dominant loss mechanism. The data are indicative of a strongly spin–dependent electron mean free path. It is suggested that the spin–polarization enhancement and the surface sensitivity of spin–polarized secondary–electron spectroscopy is due to these spin–dependent inelastic processes. The temperature dependence of the inelastic spin asymmetries is suggestive of a local band behavior, i.e. a temperature-independent exchange splitting. Above T_C inelastic spin flip–scattering, as evidenced by a large depolarization, is still very strong.

VII.1. Introduction

The problem of spin–dependent scattering mechanisms and the possibility of a spin–dependent electron mean–free–path in ferromagnetic materials and the possible influence of spin–dependent scattering mechanisms on e.g. spin–polarized photoemission spectrocopy has been of interest for some time [VII.1]. Only recently has it become possible to experimentally study the spin–dependent scattering mechanisms in detail by scattering of a monochromatic spin–polarized electron beam off a ferromagnetic surface and analyzing the energy and spin-polarization as a function of energy loss. Venus and Kirschner [VII.2] and Dodt *et al.* [VII.3] performed experiments on bulk Fe single–crystal surfaces and epitaxial Fe films, respectively, and showed that exchange scattering plays a significant role in the energy–loss process. Iron, because of its large exchange splitting ($\sim$2 eV) is a favorable case experimentally since it requires only modest energy resolution ($\leq$ 0.5 eV). On the other hand, Ni as a saturated or "strong" ferromagnet, i.e. the majority spin d–band completely filled and only minority holes available above E_F, should greatly facilitate the theoretical interpretation of spin polarized energy loss data. The small exchange splitting of Ni ($\sim$0.3 eV) requires the use of monochromatized electron beams in order to achieve the required resolution ($\leq$100 eV).

VII.2. Experiment

The polarized electron beam is derived from a negative electron–affinity GaAs
(100) crystal using circularly polarized light from a GaAlAs laser diode. The
beam is converted from longitudinal to transverse spin–polarization by a 90°
electrostatic deflector and monochromatized by a 180° hemispherical deflector.
The scattered electrons are energy analyzed by a hemispherical analyzer and the
spin–polarization is measured by a high–energy (100–120 KV) Mott detector.
The sample is a Ni "picture frame" single crystal exposing the (110) surface.
The total scattering angle in the experiment is fixed at 90°. The sample can be
rotated about its axis so that specular or various degree of off–specular spectra
can be taken.

VII.3. Results

The analysis of a spin–polarized energy loss spectrum is shown in Fig. VII.1.
The spectrum was taken at 20° off–specular. The primary energy was 12 eV and
the energy resolution 80 meV. The sample was at room temperature ($\sim 0.5 T_C$).
Shown are the individual rates (non–flip and flip for spin–up and spin down,
respectively) [VII.4], normalized to the elastic intensity. The *elastic* scattering
is purely non–flip. For these scattering conditions the spin–up scattering is
slightly larger than for spin–down [VII.5]. In the *inelastic* regime the situation
reverses and the non–flip down rate is significantly larger than the non–flip up
rate. The flip–down rate rapidly emerges in the inelastic regime and is actually
the largest rate around 300 meV energy loss. The flip–up rate is by far the
smallest rate throughout. Fig. VII.2 shows a qualitative explanation of the
results in terms of the relevant d–band structure around E_F. The left panel
shows the exchange scattering mechanism observed as a flip–down event. The

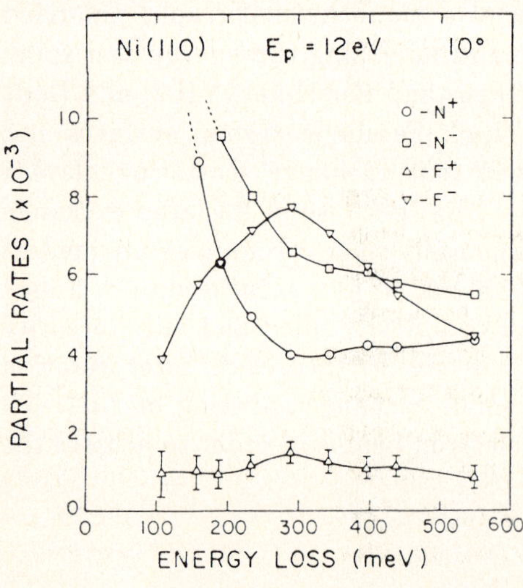

Figure VII.1. Non–flip ($N^{\pm}$) and flip
($F^{\pm}$) partial rates normalized to the
elastic intensity.

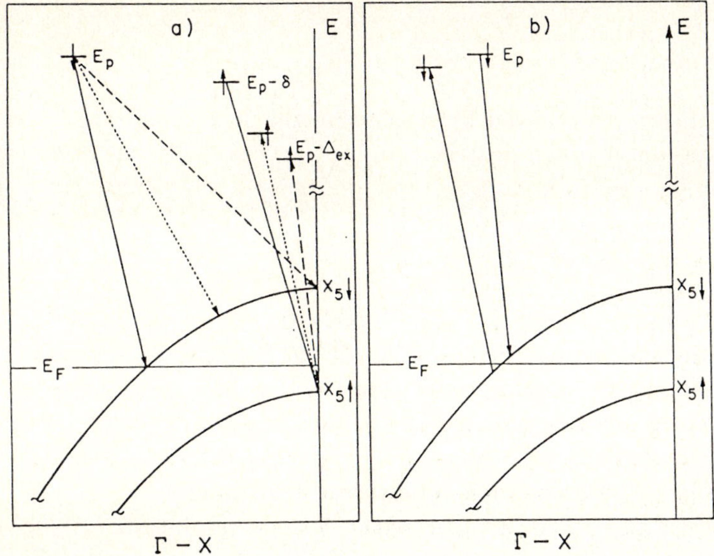

Figure VII.2. Schematic of exchange scattering processes for incoming spin–down electrons.

incoming spin–down electron falls into the empty part of the minority–spin d–band and a spin–up electron is ejected from majority band below E_F [VII.6]. Even though the experiment is angle resolved, in surface scattering only $q_\parallel$ needs to be conserved. Therefore non–$q_\perp$–conserving transitions, as indicated in Fig. VII.2, become possible. This might be at least part of the reason that we do not observe a significant angle dependence of the loss spectra (spectra at 10° and 40° off–specular are the same as the 20° spectrum). The minimum energy loss required for the flip–down process corresponds to the Stoner gap (δ), i.e. the energy of the top of the majority d–band below E_F . When taking only the d–bands into account, there is no flip–up process possible in Ni. The observed small flip–up rate can be attributed to a large extend to finite temperature transverse magnetic fluctuations. These make transitions in the "wrong" direction possible (see below). There might also be small contributions involving free–electron–like sp–states [VII.7]. Low–temperature data are required to solve this question.

The right panel of Fig. VII.2 shows the exchange contribution to the non–flip down rate. Again this process is missing for spin–up electrons, thereby explaining that $N^- > N^+$. In addition there is also the direct scattering events involving small energy excitations within the minority d–band (and sp–band) which contributes to both N^+ and N^- . From the individual rates one can obtain a spin asymmetry:

$$A = \frac{N^+ + F^+ - N^- - F^-}{N^+ + F^+ + N^- + F^-} \quad .$$

This quantity can be directly measured by measuring the scattering intensities $I_\uparrow$, $I_\downarrow$ for incoming spin–up or down polarization, respectively. The asymmetry A is given by:

$$A = \frac{1}{|P_0|} \frac{I_\uparrow - I_\downarrow}{I_\uparrow + I_\downarrow} \quad .$$

Since $N^- > N^+$ and $F^- > F^+$ (Fig. VII.1) in the inelastic regime, A is negative, and contains contributions from both the difference in the flip and non–flip rates. The direct measurement of the asymmetries does not require measuring the polarization. Therefore these measurements can be performed with much better energy resolution due to large gain in detector efficiency. In the experiment the Mott detector was replaced by a channeltron electron multiplier.

Fig. VII.3 shows asymmetry data taken with 25 meV resolution (17 meV in the inset). As expected and shown previously [VII.8] the asymmetries are negative, meaning that spin–down electrons have a higher energy loss probability than spin–up electrons. Around 300 meV energy loss the observed asymmetries are -55%. Taking into account that the data were taken at room temperature the low–temperature asymmetries can be expected to be on the order of -60, -65%. This means that the probability for loosing ~ 300 meV is ~ 4.5 times larger for spin–down electrons than for spin–up electrons!

The first rapid emergence of the asymmetry when going out of the elastic beam is due to the difference in the non–flip rates. The second rise around 65 meV energy loss (visible in the high resolution data in the inset in Fig. VII.3) can be attributed to the onset of flip–down transitions, i.e. this energy corresponds to the Stoner gap. This value is in the range (20–100 meV) previously determined by photoemision [VII.9].

VII.4. Dipole *vs.* Impact Scattering

All the results shown so far were taken well out of the specular beam. Close to specular geometry the inelastic scattering is in general dominated by so–called dipole scattering. The electron undergoes a small–angle inelastic scattering due to the long–range electric fields above the surfaces. In addition the electron is elastically reflected, either before or after energy loss. Therefore the inelastic intensity is expected to be proportional to the elastic (LEED) intensity and since the long–range Coulomb interactions do not involve the spin explicity, spin effects are expected to vanish in specular geometry [VII.10]. Fig. VII.4 shows a series of asymmetry spectra in specular geometry as a function of primary energy. The dramatic decrease of the asymmetry values when going from 16.5 eV to 19 eV correlates with the LEED intensity. At 16.5 eV the elastic intensity is at a minimum while at 19 eV it is close to a Bragg maximum, making dipole

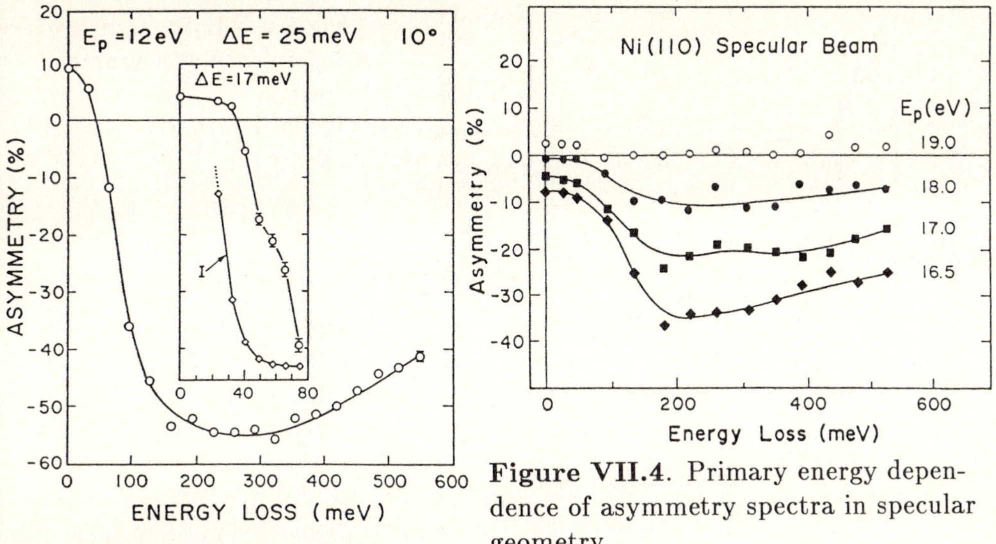

Figure VII.3. High–resolution asymmetry spectra.

Figure VII.4. Primary energy dependence of asymmetry spectra in specular geometry.

scattering the dominant loss mechanism. The spin asymmetries at this energy can be recovered by going off specular.

VII.5. Temperature Dependence

The temperature dependence of the electronic structure of itinerant magnets is of great current interest. Especially for Ni the situation is unclear due to the small exchange splitting. Photoemission and inverse photoemission show a merging of the spin–split peaks with increasing temperature although the behavior may depend on wave vector [VII.11]. Since the spin asymmetries are dominated by the difference of the flip rates and the asymmetry spectrum gives information on the average exchange splitting, e.g. a reduction in the exchange splitting should be visible as a change in the shape of the asymmetry curve, with the weight shifting to lower loss energies. Fig. VII.5 shows a series of asymmetry spectra as a function of temperature. No change in the shape of the asymmetry spectra is discernible. This finding is consistent with a local–band behavior, i.e. a temperature independent exchange splitting, but contrary to the conclusions from photoemission results. Our results are consistent with the conclusions from a recent independent similar study [VII.12]. Polarization analysis shows that even above T_C inelastic spin–flip scattering is still very strong. This is evidenced from a strong depolarization at 300 eV energy loss [VII.13].

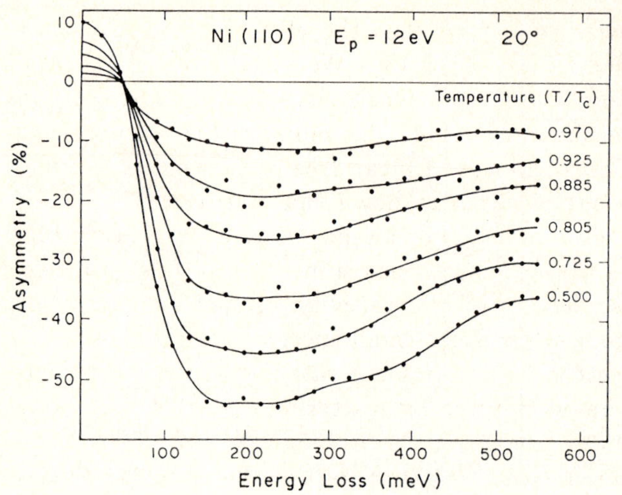

Figure VII.5. Temperature dependence of asymmetry spectra.

VII.6. Effects of Spin–Dependent Scattering on Secondary–Electron Polarization

The spin–polarization of low–energy secondary electrons shows two very surprising effects. First, the spin–polarization at very low energies is greatly enhanced (2–3 times) compared to the average band polarization. This has been established for a number of cases involving $3d$ metals [VII.14]. Second, the spin–polarization at low energies is much more surface sensitive than expected from the "universal" electron mean–free–path curve. Fig. VII.6 shows the temperature dependence of the spin–polarization at two different energies (0 eV and

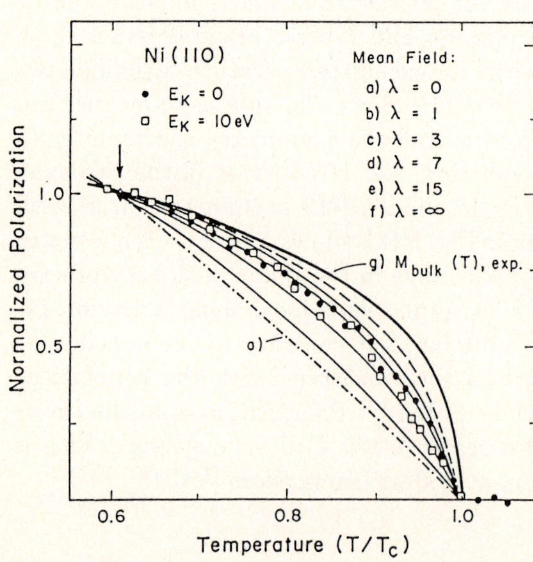

Figure VII.6. Temperature dependence of spin–polarization of low–energy secondary electrons compared to model calculations based on a layer–dependent mean–field approximation to extract effective magnetic probing depths.

10 eV). From these data it was concluded that the effective magnetic probing depth is only on the order of 5 Å [VII.15]. We believe that the reason for these effects —enhancement and surface sensitivity— can be found in the strong spin dependence of inelastic scattering. The enhancement comes about by the combination of two effects. First, the mean free path becomes strongly spin–dependent, being much smaller for spin–down than for spin–up. This is due to both the difference in the non–flip and the flip rates. Second, the strong flip–down scattering increases the number of hot spin–up electrons at the expense of spin–down electrons. The surface sensitivity can then be explained by the dominance of exchange scattering for small energy losses. Even though the low–energy secondary electrons may originate rather deep in the sample, their spin polarization could be determined by scattering in the last few layers before escaping into the vacuum. Therefore the probing depth should be identified with an "exchange mean–free–path" instead of the conventional inelastic mean free path. It would be highly interesting to perform model calculations on the secondary cascade using the inelastic scattering rates of Fig. VII.1 as input parameters [VII.16].

This work has been supported by NSF Grant No. DMR 86–00668.

References

VII.1 U. Banninger, G. Buch, M. Campagna, and H. C. Siegmann, J. Phys. (Paris) Colloq. **32**, 1 (1971); K. Sattler and H. C. Siegmann, Phys. Rev. Lett. **29**, 1565 (1972).

VII.2 A. Venus and J. Kirschner, Phys. Rev. B **37**, 2199 (1988).

VII.3 Th. Dodt, R. Rochow, D. Tillmann, and E. Kister, *Proceedings of the 12th International Colloquium on Magnetic Films and Surfaces*, Le Cruesot, France 1988; Th. Dodt, Ph, D. Thesis, Universität Düsseldorf (1988), unpublished.

VII.4 See e.g. H. Hopster, D. L. Abraham, and D. P. Pappas, J. Appl. Phys. **64**, 5927 (1988), or A. Venus and J. Kirschner (Ref. 2) for a derivation of the individual rates from the experimental data.

VII.5 D. L. Abraham and H. Hopster, Phys. Rev. Lett. **59**, 2333 (1987).

VII.6 S. Yin and E. Tosatti, International Centre for theoretical Physics, Trieste, Report No. 1C/81/129, 1981 (unpublished).

VII.7 D. R. Penn, Phys. Rev. B **35** 1910 (1987).

VII.8 J. Kirschner, D. Rebenstorff, and H. Ibach, Phys. Rev. Lett. **53**, 698 (1984).

VII.9 W. Eib and S. F. Alvarado, Phys. Rev. Lett. **37**, 444 (1976); D. E. Eastman, F. J. Himpsel, and J. A. Knapp, Phys. Rev. Lett. **40** 1514 (1978); R. Raue, H. Hopster, and R. Clauberg, Phys. Rev. Lett. **50**, 1923 (1983).

VII.10 There may be residual effects due to the spin dependence of the elastic scattering D. L. Mills, Phys. Rev. B **34**, 6099 (1986).

VII.11 J. Kanamori in *Core Level Spectroscopy in Condensed Systems*, Series in Solid–State Sciences, Springer Verlag (Berlin 1987); H. Hasegawa, Chapter XX, these Proceedings.

VII.12 J. Kirschner and E. Langenbach, Solid State Commun. **66**, 761 (1988).

VII.13 D. L. Abraham, Ph. D. Thesis, University of California, Irvine (1988) (unpublished).

VII.14 See e.g. M. Landolt in *Polarized Electrons in Surface Physics*, ed. by R. Feder, World Scientific (Singapore 1985).

VII.15 D. L. Abraham and H. Hopster, Phys. Rev. Lett. **58**, 1352 (1987).

VII.16 J. Glazer, Ph. D. Thesis, International School for Advanced Studies, Trieste (1984).

VIII. Absolute Determination of the Magnetic Anisotropy of Ultrathin Gd and Ni Films on W(110)

A. Berghaus, M. Farle, Yi Li, and K. Baberschke

Institut für Experimentalphysik, Freie Universität Berlin,
D-1000 Berlin 33, Germany

The Ferromagnetic Resonance of 130 Å and 1300 Å Gd films on a W(110) substrate is measured as a function of temperature and orientation of the magnetic field with respect to the film plane. The magnetic anisotropy energies of the 130 Å and 1300 Å films are enhanced by a factor of ten and a factor of two compared to bulk Gd. The first– and second–order uniaxial anisotropy constants K_2 and K_4 are determined. Both constants normalized with respect to the demagnetization energy $2\pi M^2$ determine the easy axis of magnetization. A theoretical discussion for this dependence is given. The detectability and interplay of K_2 and K_4 is also analyzed for the FMR of a 20 Å Ni film.

VIII.1. Introduction

The Magnetic Resonance technique has a longstanding tradition for the investigation of the magnetic anisotropy of thin films and interlayers. Most of the experiments have been performed deep in the ferromagnetic phase, that is to say, the ferromagnetic resonance (FMR). Only few experiments did investigate the vicinity of the phase transition, at T_C and above, that is to say the paramagnetic resonance (EPR). Traditionally, the films were prepared in a moderate vacuum ($\sim 10^{-8}$ mbar), covered by a nonmagnetic layer and then measured in conventional resonance spectrometers in laboratory air. A recent example is the study of Mo–Ni superlattices and its anisotropy energy [VIII.1]. Chappert *et al.* [VIII.2] did evaporate Co–monolayers on a polycrystalline Au substrate in clean 10^{-10} mbar vacuum but still a Au cover film was needed to protect the film during the resonance experiment. Only very recently the UHV in situ ESR became available [VIII.3–VIII.5]. We have shown for 0.8, 1.6, 2.8 monolayers (ML) Gd that the film behaves as a two–dimensional Ising system and changes into a three–dimensional system for 80 Å and thicker films [VIII.6].

In the present contribution we investigate Gd/W(110) films of several hundreds Ångstroms thickness in the ferromagnetic phase and show how the magnetic anisotropy energy changes between bulk material and thin films. Most of the other experimental techniques determine an effective anisotropy constant and some type of effective surface anisotropy. We will follow the well established definitions of solid state physics and use the first– and second–order terms K_2 and K_4 for the uniaxial anisotropy [VIII.7]. It is shown for some recent FMR data of $\approx$ 20 Å Ni films that the magnetic resonance is a very sensitive technique to determine both parameters. For the Gd films we present the first "complete FMR experiment", namely a temperature dependent and angular dependent experiment. The independent data sets yield the same anisotropy constants within the experimental error bars.

VIII.2. Theoretical Model

The coordinate system used in our calculations is shown in Fig. VIII.1. The free energy density of an hexagonal system such as our Gd(0001) films on W(110) is then given by the expression [VIII.2]

$$E = -H \cdot M(H,T) \sin\theta \cos(\phi_H - \phi) + 2\pi M(H,T)^2 \sin^2\theta \sin^2\phi$$

$$-[K_2(T) + 2K_4(T)] \sin^2\theta \sin^2\phi + K_4(T) \sin^4\theta \sin^4\phi \qquad (VIII.1)$$

where the first term represents the Zeeman energy, the second the demagnetization energy of a thin film and the last two terms are due to the uniaxial anisotropy energy with the c–axis parallel to the y–axis. $K_2(T)$ and $K_4(T)$ are the temperature dependent first– and second–order anisotropy constants [VIII.7]. Their effective values in thin films depend on the volume magnetocrystalline anisotropy and the stress anisotropy. The surface anisotropy which is also considered in the literature is not important in our case since our films grow in the Stransky–Krastanov mode (monolayer formation with subsequent three-dimensional island formation) [VIII.8]. Since the dc magnetic field H is applied only in the xy plane in our resonance experiment, the equilibrium position of the magnetization is given by $\theta_{eq} = \pi/2$ and for ϕ_{eq} by the relation $\delta E/\delta\phi = 0$:

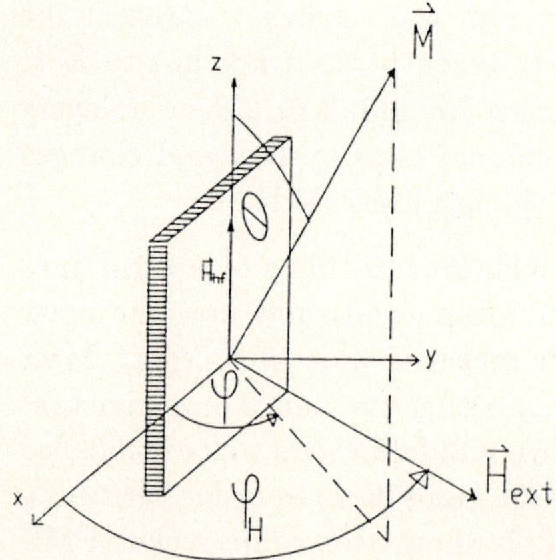

Figure VIII.1. Orientation of the dc magnetic field H and of the magnetization $M(H,T)$ with respect to the coordinate system used in our calculations. The microwave field lies always in the film plane. The dc–field H can only be rotated in the xy plane. In uniaxial symmetry along y (i.e. the hexagonal Gd c–axis) one has $\theta_{eq} = \pi/2$.

$$M(H,T)\cdot H\sin(\phi_H - \phi_{eq}) = 2[2\pi M(H,T)^2 - K_2(T) - 2K_4(T)]\sin\phi_{eq}\cos\phi_{eq}$$

$$+4K_4(T)\sin^3\phi_{eq}\cos\phi_{eq} \ . \qquad\qquad (VIII.2)$$

In zero external field ($H=0$ G) the left–hand side of Eq. VIII.2 vanishes and the easy direction of the spontaneous magnetization is determined by the competition of the demagnetization energy ($2\pi M^2$), which favors alignment in the film plane, and the magnetocrystalline anisotropy energy (K_2, K_4), which favors a perpendicular orientation for a positive K_2 (Fig. VIII.2). There exist three solutions to the free energy minimum as determined from Eq. VIII.2:

a) $\phi_{eq} = 0°$ (small magnetic anisotropy)

b) $\phi_{eq} = 90°$ (very large positive value of K_2) $(VIII.3)$

c) $\cos^2\phi_{eq} = \dfrac{2\pi M^2 - K^2}{2K_4}$ (large positive value of K_4)

The easy direction of M given by the relative values of K_2 and K_4 with respect to $2\pi M^2$ is shown in Fig. VIII.2. Besides the trivial cases of in–plane ($\phi_{eq} = 0°$)

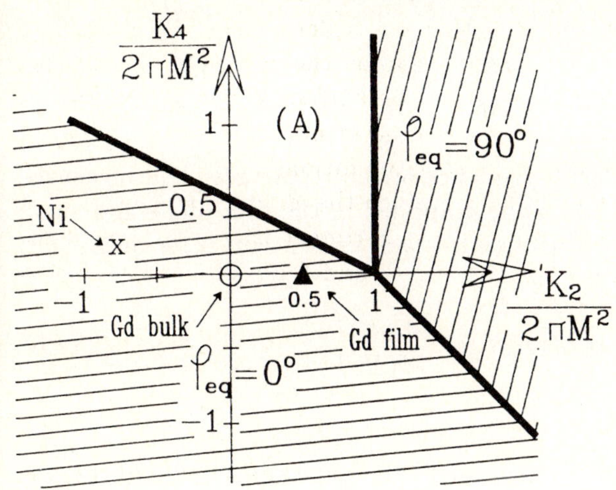

Figure VIII.2. Stability regions of the easy direction of the spontaneous magnetization as a function of M, K_2, K_4 according to Eq. (VIII.2). The hatched areas indicate easy directions parallel ($\phi_{eq} = 0°$) and perpendicular ($\phi_{eq} = 90°$) to the film plane. For the open area (A) there is a solution of a conical structure around the film normal, i.e. the c–axis with ϕ_{eq} given by $K_4\cos^2\phi_{eq} = 2\pi M^2 - K_2$. The K_2, K_4 "coordinates" of the 20 Å Ni(111) film (x), of the Gd(0001) films ($\triangle$) and of Gd bulk (O) are indicated (compare Table VIII.1).

and perpendicular ($\phi_{eq} = 90°$) anisotropy due to the dominance of either dipolar effects ($2\pi M^2$) or spin–orbit induced anisotropy (K_2, K_4) one finds a range of K_2 and K_4 values which yield a conical structure of the magnetization with respect to the film normal (Eq. VIII.3c). It is worthwhile to note that a conical structure can only be determined in the experiment if one considers also a second–order constant in the analysis of the experimental data.

The resonance–field H_R which is measured in an FMR–Experiment with the magnetic energy given by Eq. VIII.1 is calculated from the general equation [VIII.9]

$$\left[\frac{\omega_0}{\gamma}\right]^2 = \frac{1}{M^2 \sin^2 \theta}\left[\frac{\partial^2}{\partial \theta^2}E \cdot \frac{\partial^2}{\partial \phi^2}E - \left(\frac{\partial^2}{\partial \theta \partial \phi}E\right)^2\right] \qquad (VIII.4)$$

which is to be evaluated at the equilibrium position θ_{eq} and ϕ_{eq}. We obtain the following expression:

$$\left[\frac{\omega}{\gamma}\right]^2 = \left[H\cos(\phi_H - \phi_{eq}) + \right.$$

$$\left. (4\pi M - \frac{2K_2}{M} - \frac{4K_4}{M})\cos 2\phi_{eq} + \frac{4K_4}{M}(3\sin^2 \phi_{eq} - \sin^4 \phi_{eq})\right]$$

$$\cdot \left[H\cos(\phi_H - \phi_{eq}) + (4\pi M - \frac{2K_2}{M} - \frac{4K_4}{M})\sin^2 \phi_{eq} + \frac{4K_4}{M}\sin^4 \phi_{eq}\right].$$

$$(VIII.5)$$

Eq. VIII.5 determines the angular dependence of the resonance–field measured in our experimental set up. For the dc field applied in the film plane and small magnetic anisotropy Eq. VIII.5 reduces to the well known Kittel formula

$$\left[\frac{\omega}{\gamma}\right]^2 = H \cdot \left[H + 4\pi M(H,T) - \frac{2K_{\mathrm{eff}}(T)}{M(H,T)}\right] \qquad (VIII.6)$$

with $K_{\mathrm{eff}}(T) = K_2(T) + 2K_4(T)$.

It is evident from Eq. VIII.6 that in an FMR–Experiment, where the angle of the dc–magnetic field is fixed at $\phi_H = 0°$, only an effective anisotropy constant $K_{\mathrm{eff}}(T)$ is determined, provided the field and temperature dependent magnetization $M(H,T)$ is known. To obtain more detailed information on the first– and second–order contributions (K_2 and K_4) to the magnetic anisotropy an resonance–field measurement is needed. Using Eq. VIII.5 for the resonance–field values as a function of the angle ϕ_H yields the anisotropy constants K_2 and K_4. This is described in more detail in Section VIII.3.

In the following Sections we will present results for Gd(0001) films which were obtained a) as a function of temperature down to ≈ 100 K in the parallel magnetic field configuration and b) as a function of the angle ϕ_H at a fixed temperature T.

VIII.3. Experimental Results

a) Temperature Dependence $H_R(T)$, $\phi_H = 0°$

The experimental apparatus which we use to perform our FMR/EPR measurements in situ in UHV has been described earlier [VIII.4]. Gd(0001) films with thicknesses from 1300 Å down to 0.8 monolayer (ML) were evaporated on W(110) and characterized by LEED and Auger spectroscopy. Magnetic resonance at 9 GHz is performed in situ on freshly evaporated films in UHV ($p \leq 5 \times 10^{-11}$ mbar base pressure). Results for 0.8 ML, 1.6 ML, 2.8 ML and 80 Å in the vicinity of the Curie temperature have been reported earlier. A reduced Curie temperature with respect to the bulk ($T_C = 292.5$ K) was determined for all epitaxial films, e.g. $T_C(0.8ML) = 271 \pm 2$ K [VIII.6], and a strongly enhanced uniaxial magnetic anisotropy energy was found in the vicinity of the respective Curie temperatures [VIII.10]. Some representative values of $K_{eff}(T)$, as determined in [VIII.10] are included in Table VIII.1 for easier reference. No resonance–field data have been taken for these films. In the present contribution we focus on the thicker 70, 130 and 1300 Å Gd(0001) films on W(110) and on one recently prepared 20 Å $fcc(111)$ Ni film on W (110). These data illustrate the usefulness of our UHV–magnetic resonance technique to determine details of the magnetic anisotropy deep in the ferromagnetic phase.

In Fig. VIII. 3 we show typical FMR spectra of a 130 Å Gd(0001) film at different temperatures. Two characteristic features of the resonance line as a function of temperature are noticed. The line intensity rises sharply in the vicinity of the bulk Curie temperature and the resonance position shifts to lower magnetic fields in the field configuration $\phi_H = 0°$ used in this experiment. The first effect is due to the increase of the magnetic susceptibility near T_C as we have shown in Ref. [VIII.6]. The shift of the resonance–field below T_C is caused by the increase of the magnetization. The effect of the magnetic anisotropy on the field shift is discussed in the following. If one assumes zero or constant anisotropy K_{eff} in Eq. VIII.6, one should observe a monotonic shift of the resonance field H_R to lower fields in all spectra. This is not the case in Fig. VIII.3. Below ≈ 240 K the field shift reverses its direction and the resonance shifts to higher fields again for decreasing temperature. This can only be explained by an increasing positive effective anisotropy $K_{eff}(T)$ in Eq. VIII.6.

To extract $K_{eff}(T)$ from the experimental resonance fields, we assume the known field and temperature dependent magnetization $M(H,T)$ of the bulk [VIII.11]. The demagnetization factors of ultrathin films are $N_\perp = 1$ (normal to the film) and $N_\parallel = 0$ (in the film plane). At that point we note that the demagnetization factor N of a hexagonal monolayer is reduced to $N_\perp = 0.94$, as we have shown earlier [VIII.10]. The g–value of bulk Gd $g=1.973$ is used, and

Table VIII.1. Typical experiment uniaxial anisotropy constants of Gd and Ni films. The values are given at the relative temperature T/T_C with respect to the reduced Curie temperatures of the thinner films. $K_{\text{eff}} = K_2 + K_4$ is determined from measurements with the dc magnetic field applied in the film plane ($\phi_H = 0$). Details of the analysis for the 0.8 ML–80 Å Gd films can be found in [VIII.10]. The analysis of the angular dependence of $H_{\text{Res}}(\phi_H)$ yields the first and second order anisotropy constants K_2 and K_4 with a very high accuracy. Bulk values of Gd are taken from Ref. [VIII.7]. The known bulk magnetization $M(H,T)$ [VIII.11] at the respective relative temperature T/T_C and resonance field is used to calculate the anisotropy energies.

Sample		T/T_C	K_{eff} 10^6 erg/cm³	K_2 10^6 erg/cm³	K_4 10^6 erg/cm³
Gd	0.8 ML	0.93	3.9	—	—
	1.6	0.93	4.6	—	—
	2.8 ML	0.93	5.2	—	—
	70 Å	0.93	1.5	—	—
	80 Å	0.93	1.0	—	—
	130 Å	0.93	1.4	1.6	±0.2
		0.38	13.0	13.0	±0.2
	1300 Å	0.93	−1.0	−1.0	±0.2
		0.38	+3.0	+3.0	±0.2
	bulk	0.93	0.05	0.01	0.03
		0.38	+1.5	−0.9	+1.2
Ni	20 Å	0.58	−0.28	−0.63	+0.175
error	bars		±7%	±4%	±4%

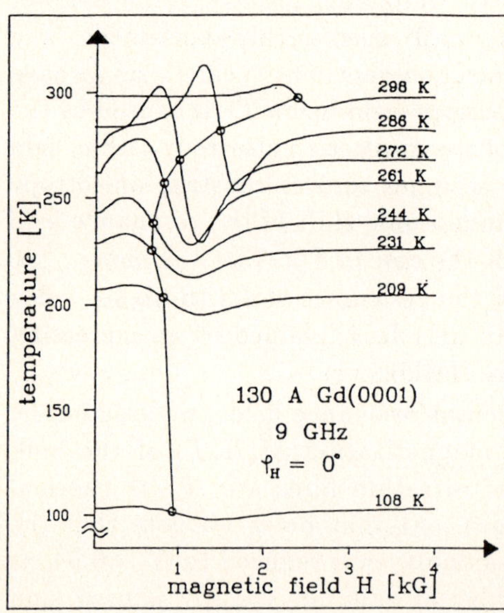

Figure VIII.3. Typical experimental FMR–spectra of 130 Å Gd(0001) for different temperatures. The *dc* magnetic field is applied in the film plane ($\phi_H = 0°$). Note the strong increase in line intensity and the shift to lower resonance fields near the Curie temperature.

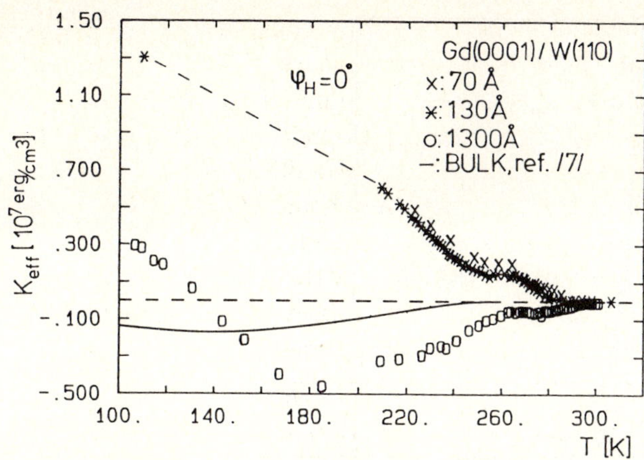

Figure VIII.4. The effective uniaxial anisotropy energy of three different Gd(0001) films as determined from Eq. (VIII.6) (see text). Note the difference in the temperature dependence of the bulk-like 1300 Å film and the thinner films which were evaporated at a lower evaporation rate. The *dc*–field is applied in the film plane for these measurements.

from Eq. VIII.6 we readily determine the effective anisotropy constant $K_{eff}(T)$ in absolute units, which is shown in Fig. VIII.4. For comparison we also plot the bulk $K_{eff}(T) = K_2(T) + 2K_4(T)$ from the literature [VIII.7]. The anisotropy of the 70 Å and 130 Å shows a different temperature dependence than the thicker 1300 Å film. $K_{eff}(T)$ of the 1300 Å film is similar to the bulk behaviour and only by about a factor of two increased. It changes sign at about 135 K. The anisotropy energy of the 70 Å and 130 Å film on the other hand is always positive and increases monotonically with decreasing temperature. Large positive $K_{eff}(T)$ favors alignment of the magnetization normal to the film plane (Fig. VIII.2). The question then immediately arises if this strongly positive anisotropy is sufficient to overcome the demagnetization energy and to force the easy direction normal to the film plane. Comparison with the demagnetization energy $2\pi M^2(T)$ of the spontaneous magnetization of bulk Gd shows that **the easy direction of all films lies always in the film plane $\phi_{eq} = 0°$ for all temperatures.** This is opposite to the known bulk behaviour, where the magnetization is parallel to the hexagonal c–axis above 240 K and tilts away from it by as much as 60° at 180 K and 40° at 110 K [VIII.12]. The reason for the in–plane magnetization of thin films is the demagnetization energy which dominates the intrinsic magnetocrystalline effects.

b) Angular Dependence $H_R(\phi_H)$, $T =$const.

As stated earlier, one can distinguish the first– and second–order anisotropy constants K_2 and K_4 only in a measurement where the resonance–field is determined as a function of the angle ϕ_H between the magnetic field and the film

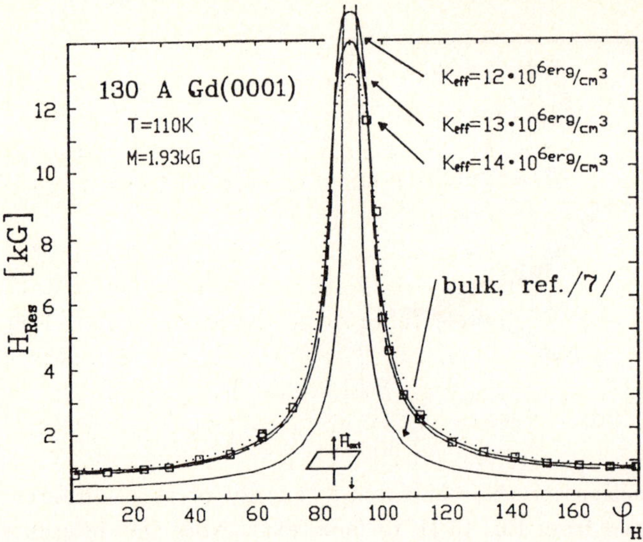

Figure VIII.5. The resonance–field H_R as a function of the angle ϕ_H between dc–magnetic field and the film plane (Fig. VIII.1) for 130 Å Gd(0001) at T=110 K. The curves are calculated according to Eq. (VIII.5). The only fit parameters entering these calculations are K_2 and K_4.

plane. Measurements of this type were performed at several fixed temperatures for a 130 Å Gd(0001) and an 20 Å Ni (111) (approx. 10 ML) film on W(110). The results at T/T_C =0.38 for 130 Å Gd(0001) are shown in Fig. VIII.5. When turning the magnetic field out of the film plane, starting at $\phi_H = 0°$, the resonance shifts to higher fields according to Eq. VIII.5. The resonance at $\phi_H = 90°$ could not be detected, since the available magnetic field strength of our magnet was limited to $\approx$ 11.5 kG. The experimental data are fitted with calculated curves for $H_{Res}(\phi_H)$ according to Eq. VIII.5. The equilibrium angle ϕ_{eq} of the magnetization entering the resonance condition (Eq. VIII.5) is numerically determined from Eq. VIII.2 at the respective resonance–field and the respective angle ϕ_H. Values for the magnetization of the bulk and the g–value are taken from the literature as before. K_2 and K_4 are the only fit parameters which enter this calculation. The best fit to the experimental data of the 130 Å Gd film at T=110 K yields $K_2 = 13 \times 10^6$ erg/cm^3 and K_4=0. This value agrees very well with the one determined in the temperature dependent measurement (Fig. VIII.4). In addition, the angular dependence reveals that the uniaxial anisotropy is only first order for the epitaxial film. This result implies that there is no possibility to obtain a conical orientation for the magnetization, since this would require a second–order contribution (Fig. VIII.2). A similar result is found at 271 K (T/T_C =0.93, Table VIII.1), where also a vanishing second–order contribution is found. From the experimental point of view one may criticize the lack of data at 90°, which impedes the determination of K_2

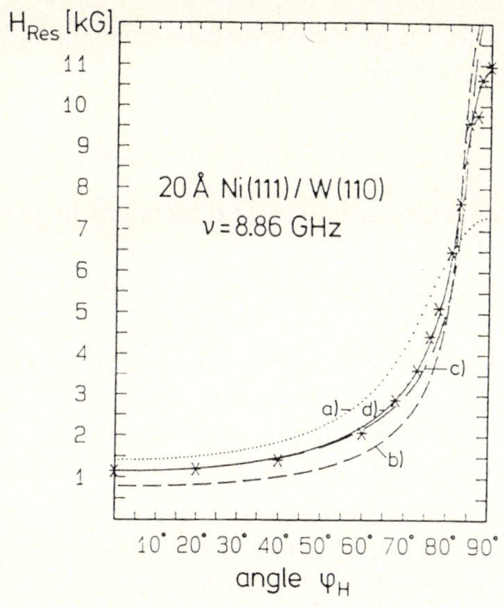

Figure VIII.6. The resonance–field H_R as a function of the *dc* field orientation (ϕ_H) with respect to the film plane for 20 Å Ni (111) on W (110) at $T/T_c =0.58$. The reduced Curie temperature $T_c = 517 \pm 5$ K of the Ni–film determined in our FMR/EPR experiment compares very well with the one previously reported for Ni (111) on Re(0001) [VIII.13]. Theoretical curves according to Eq. (VIII.2) and Eq. (VIII.5) are calculated for M=354 G, g=2.2 and a) K_2=0, K_4=0, b) $K_2 = -9.3 \times 10^5$ erg/cm³, K_4=0, c) $K_2 = -9.3 \times 10^5$, $K_4 = 3.3 \times 10^5$ d) $K_2 = -6.3 \times 10^5$ erg/cm³, $K_4 = 1.8 \times 10^5$ erg/cm³. Note that c) and d) yield the same $K_{\text{eff}} = K_2 + K_4$ and cannot be distinguished at $\phi_H =0°$.

with much higher accuracy. But for the 20 Å Ni (111) data this problem does not exist. Due to the lower magnetic moment of Ni the resonance can also be detected at $\phi_H =90°$. The calculations for this film yield as best fit parameters $K_2 = -0.63(\pm0.03) \times 10^6$ erg/cm³ and $K_4 = 0.18(\pm0.02) \times 10^6$ erg/cm³. To prove the high reliability of this determination we also show the theoretical angular dependence for $K_2 = -0.93 \times 10^6$ erg/cm³ and $K_4 = 0.33 \times 10^6$ erg/cm³, which corresponds to the same $K_{\text{eff}} = K_2 + 2K_4$ as before. From Fig. VIII.6 it becomes evident that only the resonance–field measurement reveals the correct first– and second–order contributions to the magnetic anisotropy, namely the one given in Table VIII.1.

The values for the uniaxial anisotropy of *fcc*(111) Ni films on W(110) may be compared to the known first– and second–order cubic anisotropy constants $K_1 = -5 \times 10^4$ erg/cm³ and $K_2 - +5 \times 10^4$ erg/cm³ of bulk Ni [VIII.15] which yield an easy direction of M parallel to the [111] direction, which is normal to the film plane. The uniaxial anisotropy of the 20 Å film derived in our experiment is by one order of magnitude larger. This effect has also been reported by Bergholz *et al.* [VIII.13]. These authors attribute the measured increase in the anisotropy of the thin film to a Néel–type surface anisotropy K_s (erg/cm³), which is only determined in first order. The effect of stress in the thin film due to the lattice mismatch of the substrate to the adsorbate lattice is neglected in this picture. If one calculated the corresponding volume anisotropy of the thin film (thickness d=20 Å) according to K_{eff} (erg/cm³) $= 1/d\, K_s$ (erg/cm³ with $K_s = -0.19$

erg/cm³ [VIII.16] one gets $K_{\text{eff}} = -9.4 \times 10^5$ erg/cm³, which is only slightly higher than the value determined in our experiment.

VIII.4. Conclusion

We have presented the first complete temperature and angular dependent in situ FMR measurement on well characterized Gd(0001) films in UHV. It is shown that the FMR measured with the magnetic field applied parallel to the film plane determines the effective value of the magnetic anisotropy in absolute units. More detailed information on first– and second–order contributions to the anisotropy are obtained from resonance–field measurements with very high accuracy. The easy direction of the magnetization of all Gd films investigated is found to lie in the surface plane for all temperatures measured. The temperature dependence of the 130 Å film indicates that at very low temperatures ($\approx$ 20 K) the uniaxial anisotropy favoring normal orientation of M may overcome the demagnetizing energy and the easy direction may switch to the normal orientation.

For a 20 Å Ni film at room temperature we determined the first– and second–order anisotropy constants in uniaxial symmetry. We find an negative magnetic anisotropy K_2 favoring alignment in the film plane with a magnitude by an about a factor of twenty smaller than the value for Gd. It is shown that in the case of Ni second–order contributions to the uniaxial anisotropy must be taken into account, while for Gd thin films these contributions vanish. This is of importance for theoretical calculations which determine the role of spin–orbit coupling in the magnetic anisotropy of thin films.

We have benefitted from discussions with J. Kirschner, U. Gradmann, and A.S. Arrott. This work was supported in part by Deutsche Forschungsgemeinschaft, Sonderforschungsbreich 6.

References

VIII.1 M. Y. Pechan and I.K. Schuller, Phys. Rev. Lett. **59**, 132 (1987).

VIII.2 C. Chappert, K. Le Dang, P. Beauvillain, H. Hurdequint, and D. Renard, Phys. Rev. B **34**, 3192 (1986).

VIII.3 B. Heinrich, K. B. Urquhart, A. S. Arrott, J. F. Cochran, K. Myrtle, and S. T. Purcell, Phys. Rev. Lett. **59**, 1756 (1987).

VIII.4 K. Baberschke, M. Zomack, and M. Farle *Magnetic Properties of Low–Dimensional Systems*, edited by L.M. Falicov and J.L. Morán–López, Springer Proceedings in Physics Vol. 14 (Springer, Berlin 1986) p. 84.

VIII.5 G. A. Prinz, G. T. Rado, and Y. Y. Krebs, J. Appl. Phys. **53**, 2087 (1982).

VIII.6 M. Farle and K. Baberschke, Phys, Rev. Lett. **58**, 511 (1987).

VIII.7 B. Coqblin, *The Electronic Structure of Rare–Earth Metals and Alloys: the Magnetic Heavy Rare–Earths* (Academic, London, 1977).

VIII.8 K. Baberschke, M. Farle, and M. Zomack, Appl. Phys. A **44**, 13 (1987).

VIII.9 L. Baselgia, M. Warden, F. Waldner, S. L. Hutton, J. E. Drumbeller, J. Q. He, P. E. Wigen, and M. Marysko, Phys. Rev. B **38**, 2237 (1988).

VIII.10 M. Farle, A. Berghaus, and K. Baberschke Phys. Rev. B **39**, 4838 (1989).

VIII.11 H. E. Nigh, S. Legvold, and F. H. Spedding, Phys. Rev. **132**, 1092 (1963) and M.N. Deschizeaux and G. Develey, J. Phys. (Paris) **32**, 319 (1971).

VIII.12 W. D. Corner and B. K. Tanner, J. Phys. C **9**, 627 (1976).

VIII.13 R. Bergholz and U. Gradmann, J. Magn. Magn. Mat. **45**, 389 (1984).

VIII.14 Yi Li, A. Berghaus, M. Farle, K. Baberschke to be published.

VIII.15 C. J. Gadsen and M. Heath, Solid State Commun. **20** , 951 (1976)

VIII.16 U. Gradmann, J. Magn. Magn. Mat. **54–57**, 733 (1986).

IX. Surface Magnetism of Novel Expitaxial Systems Determined by Electron Capture Spectroscopy and by Spin-Polarized Electron Emission Spectroscopy

C. Rau

Department of Physics and Rice Quantum Institute,
Rice University, P.O. Box 1892, Houston, TX 77251, USA

The ferromagnetic and critical behavior, the temperature (T), magnetic field (H), and thickness (d) dependence of magnetic anisotropies existing at surfaces of ultra–thin, epitaxial systems are determined using Electron Capture Spectroscopy (ECS) and Spin–Polarized Electron Emission Spectroscopy (SPEES). For all systems [Ni(100)/Cu(100), Ni(100)/NaCl(100), hcp Tb(0001)/W(110), Fe(100)/Ag(100), Tb/Fe(100)/Ag(100), Fe(100)/Au(100), fcc Fe(111)/Cu(111), Fe(110)/W(110), V(100)/Ag(100), Pd(100)/Ag(100), Pd/W(110)], including $Fe_{80}B_{20}$ and Ni(110), we find ferromagnetic order. We report on the T–, H– and d–dependence of magnetic anisotropies. We find that the surface Curie temperature T_{CS} depends on d. Although for V(100)/Ag(100), the measured critical exponent $\beta=0.128$ agrees with $\beta=1/8$ as given by the two–dimensional Ising model, for other systems we find disagreement with theoretical predictions.

IX.1. Introduction

Ion–surface interaction at small angles of incidence provides a powerful means to study the magnetic and electronic properties of the topmost atomic layer of magnetic systems. At *grazing* angles of incidence, ions cannot penetrate into a surface, they are specularly reflected and, therefore, interact only with the topmost surface layer. This simple fact reveals the extreme surface sensitivity of experimental methods where grazing–angle ion–surface interaction is used to retrieve data on the magnetic properties of surfaces.

Magnetic systems in the form of thin films are ideal for exploring basic concepts in theoretical physics such as phase transitions in two or three dimensions or finite–size scaling of thermodynamic quantities which lead to determine various critical exponents [IX.1,IX.2]. Furthermore, the magnetic properties of surfaces and thin films are of pivotal importance in modern magnetic storage devices.

Ultra–thin (1–30 atomic layers) magnetic films, prepared as two–dimensional or quasi–two–dimensional expitaxial structures on single crystal metal substrates are expected to exhibit unusual electronic and magnetic properties [IX.1]. Epitaxial films can be grown in such a way that atomic spacings and/or structures are not those characteristic of bulk specimens of the same material. Therefore, they can be viewed as artificial or property–modified materials [IX.3].

Springer Proceedings in Physics, Vol. 50   **Magnetic Properties of Low-Dimensional Systems II**
Editors: L.M. Falicov · F. Mejía-Lira · J.L. Morán-López © Springer-Verlag Berlin, Heidelberg 1990

The use of artificially structured materials directly permits the control and tailoring of dimensionality, symmetry and magnetic couplings [IX.4].

A very interesting case corresponds to an element which is paramagnetic in the bulk and may become ferromagnetic when its atoms are properly deposited on a substrate of a different nonmagnetic material which imposes its own lattice constant to the ultra–thin film (see Section IX.3.3).

The most interesting case of such a magnetic "enhancement" corresponds to an element which possesses *no* magnetic moment as an ensemble of isolated atoms and, however, may become ferromagnetic when its atoms are deposited on an appropriate *nonmagnetic* substrate which imposes its own lattice constant to the ultra–thin, artificially structured film (see Section IX.3.4).

For the investigation of surface magnetic structures, we use ion–induced *capture* or *emission* of spin–polarized electrons, which is a powerful means for probing various surface properties, in particular surface magnetic properties. Electron Capture Spectrocopy (ECS) and Spin–Polarized Electron Emission Spectrocopy (SPEES) permit the detection of "long–range" and "short–range" electron spin polarization (ESP) with extreme surface sensitivity [IX.5].

Using ECS, we studied the long–range ESP of *fcc* Ni(100) films on Cu(100) and NaCl(100) substrates [IX.5,IX.6]. We find that the ESP of 7 monolayer–(ML)–thin Ni(100) films on NaCl(100) agrees very well with the ESP of bulk Ni(100). Using Cu(100) as substrate for a 7 ML thin Ni(100), we find a strong reduction of the surface–ESP caused by the influence of non–polarized electrons from the Cu(100) substrate. These disturbing effects, caused by band overlap of Ni and Cu electron bands, can be avoided by using NaCl(100) as substrate.

Face–centered cubic Cu, however, is an ideal substrate to stabilize the high temperature *fcc* phase of Fe. Using ECS, we find that, at 298 K, non–zero long–range ESP exists at free (uncoated) surfaces of 4 ML *fcc* Fe(111)p(1×1) homogeneously (island–free) deposited on Cu(111). At 400 K, nonzero short–range ESP is still detected for 1 and 2 ML of *fcc* Fe(111) on Cu(111) [IX.7].

We have used Ag(100) and Au(100) —where no overlap of Fe and Ag or Fe and Au electronic bands is expected— as substrates for thin (1–10ML), *bcc* Fe(100) films [IX.8]. We find that, between 173 K and 303 K, d–dependent, *in–plane* remanent ferromagnetic order exists at uncoated surfaces of *bcc* Fe(100)/Ag(100) films. Large T–dependent *perpendicular* (out–of–plane) magnetic anisotropies exist at surfaces of Tb–coated Fe(100)/Ag(100) films. It is found that the surface Curie temperature T_{CS} depends on d. For 2 ML Fe(100)/Au(100), near T_{CS} =290.03 K, we have determined the surface critical behavior and find a critical exponent β=0.25, in disagreement with theoretical predictions [IX.9].

For thin (50 Å thick) films of *hcp* Tb(0001) on *bcc* W(110), we find [IX.10] surface enhanced magnetic order and a critical exponent β=0.348 which is consistent with β=0.35 for the anisotropic special transition predicted by Diehl and Eisenriegler [IX.11].

We have studied the system $V(100)p(1\times1)/Ag(100)$ and find that the surfaces of artificially structured, ultra–thin films order ferromagnetically at low temperatures [IX.12]. This is in contrast to bulk V itself, which is paramagnetic at all temperatures. The temperature dependence of the surface–ESP behaves like $(T_{CS} - T)^{\beta}$, with β=0.128 which is in good agreement with the exact value β=1/8 of the two–dimensional Ising model.

We have used Ag(100) substrates to deposit ultra–thin, artificially structured, epitaxial films of *fcc* Pd(100). Isolated Pd atoms possess no magnetic moment. Bulk *fcc* Pd, however, is well–known for its high paramagnetic susceptibility. Using Ag(100) allows us to "expand" the lattice constant of Pd, which could induce ferromagnetic order in the Pd films. We find that long–range ESP exists at the surface of thin Pd(100) films [IX.13]. The ESP is very sensitive to hydrogen adsorption.

Further ECS studies relate to the investigation of ultra–thin Fe(110) and Pd films deposited on W(110) substrates. From preliminary ECS experiments, there is evidence for the existence of nonzero long–range surface ESP.

Using ECS, we have studied [IX.5] the short–range surface ESP of the amorphous ferromagnet $Fe_{80}B_{20}$ and find no evidence for so–called magnetically dead layers at atomically clean surfaces. At room temperature, the short–range ESP amounts to about 55% [IX.14].

We used SPEES to study the emission of spin–polarized electrons at surfaces of thin films of amorphous $Fe_{80}B_{20}$ [IX.14]. The ESP of the emitted electrons amounts to around 15% – 20% which clearly establishes the existence of long–range ESP at these surfaces and proves that SPEES is a powerful technique for surface magnetometry [IX.14].

Studying the *energy–* and *angle–resolved* ESP of electrons emitted during ion–surface interaction from Ni(110) picture–frame crystals, we find that the electron energy distribution $n(E)$ is *strongly* different from that of electron-induced secondary–electron spectra [IX.15].

The energy and spin distributions of the electrons, originating from the *topmost* surface layer, contain valuable information about electronic and magnetic properties at Ni(110) surfaces *and* about various atomic and electronic processes taking place at surfaces. Our experimental data do not support theoretical models that are based on pure statistical electron emission [IX.16].

IX.2. Experimental

The details of ECS and SPEES are given in [IX.5,IX.17,IX.18], here we give only a brief discussion of the essential features.

ECS is based on the *capture* of one or two spin–polarized electrons during specular surface reflection of deuterons at magnetic surfaces [IX.5]. For an angle of incidence of $0.2°$, the distance of closest approach of the ions amounts to 1–2

Å. Thus, the ions probe the spin–polarized local electron densities of state at the *topmost* surface layer.

A well–collimated deuteron beam is incident at grazing angle on the sample which is magnetized using a C–shaped electromagnet. For the measurement of the H–dependence of the in–plane component of the ESP, the samples are magnetized parallel to the surface plane. Specularly reflected particles then enter a transverse electric field which spatially separates remaining deuterons from D^0 atoms and deuterium negative ions formed by one– and two–electron capture, respectively. The ion currents are measured using two Faraday cups. The D^0 atoms, however, impinge on a T-target where their nuclear polarization is determined by measuring the asymmetry in the angular distribution of ^{4}He–particles emitted in the reaction $T(d,n)^4He$. A weak magnetic field, parallel to the magnetizing field, provides a well defined quantization axis.

In–plane, long–range ESP is detected by means of one–electron capture (OEC) processes: If the in–plane ESP is nonzero, OEC results in the formation of deuterium atoms having a net ESP. This ESP is partially converted to a nuclear polarization by hyperfine interaction. Determination of the resultant nuclear polarization provides a measure of the net polarization P of the captured electrons.

Defining the ESP P along the target magnetizing field yields

$$P = \left(n^+ - n^- \right)/\left(n^+ + n^- \right)$$

where n^+ and n^- represent fractional numbers of electrons with spin moment antiparallel (majority spin electrons) and parallel (minority spin electrons), respectively, to the in–plane magnetizing field.

For the measurement of short–range ferromagnetic order, we study two-electron capture (TEC) processes $(H^+ + 2e^- = H^-)$ or $(D^+ + 2e^- = D^-)$ [IX.17]. For H^- and D^-, the only stable bound state is the $1s^2$ state. Therefore, H^- or D^- can only be formed by capture of electrons with opposite spins. At present, we use either 20 keV hydrogen ions or 150 keV deuterons for the ESP measurements. For this energy range, the characteristic length within two electrons are captured by a single ion, amounts to about 10–12 Å. Thus, TEC processes are sensitive to the short range ESP existing on atomic scale within a few atomic neighbors. It is obvious that TEC will be strongly suppressed by the presence of short–range ESP where on an atomic scale, predominantly electrons with parallel oriented spins are available for capture by a single ion. The reduction in the D^-/D^+ or H^-/H^+ ratio in the reflected beam, relative to that for a nonmagnetic target such as Cu, provides a direct measure of the short–range ESP at a magnetic surface.

In SPEES, a methodological advancement of the ECS technique, we investigate the *emission* of spin–polarized electrons which is induced during the grazing–angle ion–surface reflection. Using an *einzellens*, electrons emitted along the surface normal (emission cone angle: 11°) of a remanently magnetized target, are focused on an electrostatic energy analyzer for energy probing.

For spin analysis, the electrons are accelerated to 150 eV and hit an electron spin analyzer [IX.18] based on low–energy, diffuse scattering of electrons from a high Z target (preferably Au). For zero–ESP calibration, the Au target is replaced by an Al target. The spin–orbit interaction provides the physical basis of this spin analyzer. Using two channeletrons, positioned at equal angles to the right and the left of the direction of the incoming electron beam, electrons elastically backscattered from the Au target are detected at angles of 120°. The incoming electron beam is focused on a Au target of 2 mm in diameter. This certifies that the measured count ratio of backscatterd electrons depends only on counting statistics. The count ratio of the electrons, detected in the two channeletrons, provides a direct measure of the ESP of the electrons emitted from the magnetic surface [IX.18].

For the deposition of ultra–thin magnetic films, atomically clean and flat substrate crystals are prepared under ultra–thin vacuum in a target preparation chamber [IX.5]. The surface orientation of these substrate crystals is better than 0.01°, and is monitored by use of a precision X–ray difractometer.

The flatness of substrate surfaces on an atomic scale is achieved by thermal surface smoothening [IX.5] in hydrogen atmosphere and is characterized by using grazing–angle ion reflection [IX.5] of highly collimated (angle of divergence $<$ 0.01°) deuterons at the substrate surface. The ion reflectivity I (reflected beam intensity per solid angle/incoming beam intensity per solid angle), which is extremely sensitive to the existence of atomic steps [IX.5, IX.19], amounts to 95%, clearly establishing the atomic flatness of the surfaces. From our recent work on the atomic flatness of surfaces of single crystalline substrates using a scanning tunneling microscope (STM), there is direct evidence that the ion reflectivity I is strongly correlated to the presence of atomic steps as was already indicated in the past by our computer simulations on the intensity and energy distribution of ions in surface and subsurface channeling [IX.19].

Applying standard cleaning and annealing procedures, developed in our earlier studies, and using Auger–electron spectroscopy with a cylindrical–mirror analyzer, residual C and O contaminations are measured to be less than 1% of a monolayer. The single crystalline state of the substrate and film surfaces is detected by low– and high–energy electron diffraction (LEED and RHEED).

Ultra–thin magnetic films are deposited by electron beam evaporation at 8×10^{-10} mbar. The thickness of the films is determined with a calibrated quartz oscillator, with calibrated Auger electron signals and by using RHEED oscillations.

The island–free growth of the films is checked by monitoring the ion reflectivity I and the energy distribution of the specularly reflected deuterons. At the surface of all films, studied so far, I is not reduced from its initial value of 95%, measured at the atomically flat substrate surfaces. Furthermore, the presence of islands would yield an additional energy loss of the deuterons caused by penetration of the ions through islands by planar channeling, which is not observed. In addition, we use RHEED and RHEED oscillations to check film growth modes.

IX.3. Results and Discussion

Much theoretical work has been centered upon the electronic and magnetic properties of thin films of ferromagnetic materials. A number of interesting questions can be answered by measuring the ESP at surfaces of thin films. One may ask: Does ferromagnetism already exist in films 1 or 2 ML thick? A point of considerable interest is the investigation of magnetic anisotropies and of phase transitions in thin films and the determination of critical exponents.

IX.3.1 Surface Magnetism of Ultra–Thin, Epitaxial Films: [*fcc* Ni(100) on Cu(100) and NaCl(100), *fcc* Fe(111)/Cu(111)]

Using ECS, we have studied the magnetic order at surfaces of 2 ML Ni(100)/Cu(100) [IX.6,IX.5]. At 300 K, the in–plane ESP is nonzero which proves that ferromagnetism already exists in 2 ML Ni(100)/Cu(100). Increasing d up to 64 layers, the ESP increases from -19% for $d=2$ ML to -33% for 8 ML and up to -65% for $d=64$ ML which is close to the ESP of -64% detected at the surface of bulk Ni(100) [IX.5]. Contrary to this, the ESP of a 7 ML thick Ni(100)/NaCl(100) film amounts to -64% revealing that bulk magnetism already exists in thin films 7 ML Ni(100). The results of these experiments suggest that this effect might be caused by the influence of non–polarized Cu–substrate electrons, which, due to electron–phonon scattering, at room temperature have a mean free path of approximately 2–4 nm and, therefore, in case of overlap of electronic bands, can diffuse into the Ni(100) film thereby reducing the ESP at the Ni(100) surface.

The high temperature *fcc* phase of Fe can be stabilized by deposition of epitaxial Fe films on *fcc* Cu substrates. Recently, the magnetic properties of surface–coated *fcc* Fe films were studied, with contradictory experimental results [IX.20–IX.23].

In order to settle the key issue of ferromagnetism *versus* antiferromagnetic ordering in *fcc* Fe films, we have investigated the long–range and the short–range ESP of *fcc* Fe(111)/Cu(111) films.

At surfaces of 1–4 ML thin *fcc* Fe(111)/Cu(111) films, we find at 300 K *nonzero* long–range and short–range ESP which clearly establish that *ferromagnetism* exists at the *topmost* surface layer of homogeneous and island–free, un-coated *fcc* Fe(111) films on Cu(111) [IX.7]. These findings are in agreement with the results of Refs. IX.20 and IX.21 reported for surface–coated *fcc* Fe(111) films, and they are in contrast to earlier claims according to which surface–coated *fcc* Fe(111) films possessing an island structure were paramagnetic at 300 K and antiferromagnetic at lower temperatures [IX.22,IX.23].

IX.3.2 Magnetic Anisotropies at Surfaces of Thin Films: Uncoated and Tb–Coated *bcc* Fe(100)/Ag(100)

We have investigated the surface ESP of ultra–thin *bcc* Fe(100)/Ag(100) [IX.8]. The magnetizing field H is applied in the surface plane of the Fe(100) films along a [011] axis. The atomic flatness of the substrate and film surfaces is investigated by means of a STM and by means of grazing–angle ion–surface reflection of deuterons. The ion reflectivity I amounts to 95%, establishing the atomic flatness of the surfaces. The epitaxial growth of Fe(100) on Ag(100) is well established [IX.24,IX.25,IX.8]. The layer–by–layer growth is confirmed by RHEED oscillations.

For Fe(100)/Ag(100) films, we find, between 173 K and 303 K, that non–zero *in–plane* remanent ESP exists at the surface of all films (d=1–10 ML). The magnetic field is varied between H=0 and H=600 Oe, and a marked H–dependence of the ESP is *only* found for Fe films consisting of 1 ML. Films with d between 2 and 10 ML are already saturated in fields of a few Oe, whereas 1 ML thin films reach saturation only at about 300 Oe. This directly implies the existence of a strong thickness–dependence of the *in–plane* magnetic anisotropy.

For 1 ML thin films, at 173 K and H=600 Oe, the saturation–ESP amounts to +13%, remains constant down to H=300 Oe and decreases to a remanent–ESP of 5% for H=0. This behavior, measured along the Ag[011] direction in the Ag(100) plane can be explained by the existence of a magnetocrystalline anisotropy existing along Fe[001] and Fe[010], the direction of easy surface magnetization being at 45° to the Ag[011] directions. For H=0, the remanent ESP along a [001] easy–direction then amounts to $5\sqrt{2}$=7%.

In a further ECS experiment, we have coated a 3 ML thick Fe(100)/Ag(100) film with 2 ML Tb and find a behavior of the surface ESP, now measured at the surface of the Tb layers, that contrasts strikingly with that of uncoated Fe–surfaces. For T=137 K, the ESP, measured in the surface plane, remains zero for all applied fields (up to 900 Oe) in accordance with the existence of a strong magnetic anisotropy *perpendicular* to the surface plane. Increasing T up to 213 K, we detect, for $H > 600$ Oe, an ESP of approximately 9%. This is in agreement with the existence of a strongly temperature–dependent out–of– plane magnetic anisotropy which, for instance at 137 K, cannot possibly be overcome by magnetic fields of several kOe. Next, we keep H at 300 Oe and increase T from 303 K to 373 K. This causes the ESP to decrease from 8.5% at 303 K to zero at 373 K indicating that the surface Curie temperature T_{cs} of the Tb–(2 ML)–coated Fe/Ag films is at around 373 K.

IX.3.3 Surface Critical Behavior of Ultra–Thin Magnetic Films: V(100)/Ag(100), Tb(0001)/W(110), Fe(100)/Au(100)

Using ECS, we studied the magnetic behavior of truly two–dimensional magnetic films consisting of a few monolayers of V(100)p(1×1) on Ag(100) substrates [IX.12].

For a substrate temperature of 373 K and an evaporation rate of 0.04 Å/s, homogeneous and island–free epitaxial growth of the V films are obtained [IX.12]. The specimens are magnetized along the [001] direction in magnetic fields ranging between 30 Oe and 600 Oe. Such applied fields have a negligible effect on the ESP in the investigated temperature range [IX.26,IX.27].

For all film thicknesses (d=1–7 ML), nonzero values of the ESP are observed, clearly establishing that long–range ferromagnetic order exists at the topmost layer of the films.

Fig. IX.1 shows the T–dependence of the normalized long–range ESP $P/P_\circ$ at the surface of 5 ML thick V(100)p(1×1)/Ag(100) films as function of T/T_{cs}, with T_{cs} =475.1 K and $P_\circ = -17.8\%$ the calculated ESP at $T = 0$. In Fig. IX.2, a logarithmic plot is given of the ESP as function of the reduced temperature $(T_{cs} - T)/T_{cs}$. The critical temperature T_{cs} and the critical exponent β for a five–layer film are determined simultaneously by a linear least–square fit of

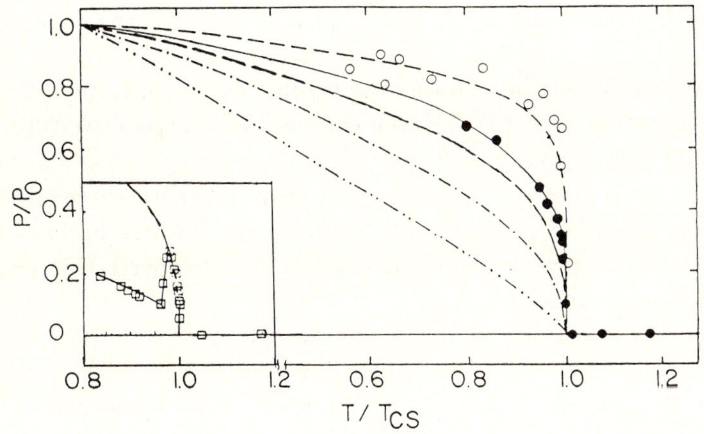

Figure IX.1. Normalized ESP $P/P_\circ$ as function of T/T_{cs} for 5 ML thin films of V(100) on Ag(100) (circles), for 50 Å–thin Tb(0001)/W((110) films (squares in the inset) and for 2 ML thick Fe(100) film on Au(100) (full circles). The lines represent the theoretical predictions for the spontaneous magnetization of the two–dimensional Ising model (– – – –), the three dimensional Ising model of the isotropic XY and Heisenberg models (— — —), the bulk mean–field model (– · – · –), the bulk mean–field model (– ·· – ·· –). The solid line represents $\beta = 0.25$.

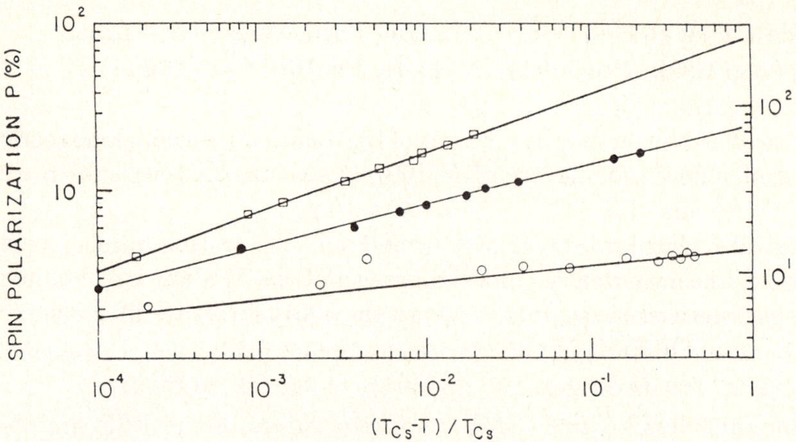

Figure IX.2. Log–log plot of the ESP data shown in Fig. IX.1 5 ML V(100)/Ag(100) (circles, P values at the right side of the Figure): $\beta = 0.128$ and $T_{cs} = 475.1$ K; 2 ML Fe(100)/Au(100) (full circles): $\beta = 0.25$ and $T_{cs} = 249.96$ K). The solid lines are drawn for $\beta = 0.125$, 0.250 and 0.350, respectively.

the polarization data under the assumption of a power law behavior of the form $(T_{cs} - T)^{\beta}$. The mean least square fits of the data were performed by varying T_{cs} in steps of 0.01° yielding a best value $T_{cs} = 475.1 \pm 0.02$ K. For that optimal value of T_{cs}, β is found to be $\beta = 0.128 \pm 0.01$ giving the slope of a straight line in the log–log representation of the P data shown Fig. IX.1. The respective curve in Fig. IX.2 corresponds to $\beta = 0.125$, which cannot be distinguished from 0.128 on the scale of the figure.

The value we obtain for the critical exponent β agrees well with the exact value 1/8 of the two–dimensional Ising model of a ferromagnet. For comparison, we also show, in Fig. IX.1, the temperature dependence of the magnetization as predicted by various theoretical models.

From measurements of the ESP as a function of d and t we find that T_c of a film decreases with d, as expected. This is particularly apparent from the fact that below 3 ML the V(100) samples have to be cooled below room temperature to exhibit ferromagnetism.

The present results are consistent with ground–state one–electron band calculations of Yokoyama *et al.* [IX.28], who studied unsupported five–layer thick V(100) films, and of Fu, Freeman, and Oguchi [IX.29], who analyzed one and two layers of V(100) on Ag(100), and of Gay and Richter [IX.30] who considered an unsupported monolayer of V(100).

For the deposition of 50 Å thin Tb(0001) films, we prepared atomically clean and flat W(110) substrate crystals. The crystals are cleaned by extensive heating cycles at 1770 K in oxygen followed flashing off at 2770 K the adsorbed oxygen which remains after removal of carbon. After further annealing and sputtering cycles at 1300 K, using AES, the residual O and C contaminations are found to be less than 1% of a monolayer.

The single–crystalline state of the *bcc* W(110) and *hcp* Tb(0001) surfaces is detected by means of LEED and RHEED. The atomic flatness of the surfaces is investigated by means of a STM and by measuring the ion reflectivity which I amounts to > 95%. For further details, see Refs. IX.31, IX.10 and IX.7.

Using ECS, P is measured in fields ranging between 25 and 600 Oe, the samples being magnetized along W[100] direction [IX.10]. In Fig. IX.1, the T-dependence of P at Tb(0001)/W(110) surfaces is given for H=250 Oe. The field was varied between 25 Oe and 600 Oe, with no observable influence on the polarization data. Nonzero P values establish that the surfaces of the films are ferromagnetically ordered up to around 250 K, which lies above both the Curie temperature $T_{Cb} = 220$ K and the Néel temperature $T_{Nb} = 228$ K. Similar behavior was recently found in ECS experiments at surfaces of polycrystalline Tb [IX.32]. Using ferromagnetic induction and Kerr effect measurements, T_{Cb} was found to be 220 K.

With increasing T, P decreases from 22% at 146 K, until it reaches a value of 7% at about 240 K, which lies slightly above T_{Cb} and T_{Nb}. As T increases further, P increases steeply to 21% at 243 K, at which it drops to zero at T_{Cs} =249.96 K.

In Fig. IX.2, the temperature variation of P in the neighborhood of T_{Cs} is shown in a log– log plot. This provides a direct determination of the critical exponent β. We obtain $\beta = 0.348 \pm 0.01$. It was recently found [IX.11] that, along an axis of easy magnetization, β can amount to 0.35. The value β=0.35 agrees well with our experimental result β=0.348 measured along an easy axis.

We note that because T_{Cs} does not strictly coincide with T_{Cb} [IX.11], one cannot fully exclude that our value $\beta = 0.348$ corresponds to an effective exponent. It appears, however, that because $(T_{Cs} - T_{Cb})/T_{Cs} \ll 1$, the predictions of Ref. [IX.11] should remain valid for the present system [IX.33].

We have further studied ultra–thin films of Fe(100) on Au(100) [IX.9]. For clean Au(100) surfaces, first (5×20) LEED patterns are obtained, which change into (1×1) patterns during Fe deposition. As for Fe(100)/Ag(100), we find, for films ranging in thickness from 1 to 3 ML, nonzero *in–plane* ESP between 113 K and 303 K. In Fig. IX.1, the T–dependence of the ESP $P(T)$ is given for d=2 ML and $H = 60$ Oe. With increasing T, the ESP decreases to zero at $T_{Cs} = 290.01 \pm 0.02$ K. Fig. IX.2 gives logarithmic plot of the ESP as function of the reduced temperature T/T_{Cs}. The determinations of T_{Cs} and β are simultaneous as described for V(100). We obtain $\beta = 0.25 \pm 0.01$. We note that the value of the critical exponent $\beta = 0.25$ does not agree with the value $\beta = 0.128$ obtained for ultrathin V(100)/Ag(100) films [IX.8], which coincides with the exact value of 1/8 of the two– dimensional, anisotropic magnetic films.

IX.3.4 Surface Magnetism of Pd(100)/Ag(100)

For Pd films in the form of multi–layers, sandwiched between Au layers, large paramagnetic susceptibilities have been observed by Brodsky and Freeman [IX.34, IX.35]. The observed effect is attributed to the expansion of the Pd lattice between the Au layers. From the observed difference between the theoretical misfit $f = 4.83\%$ and the experimentally determined average misfit $f = 1.8\%$, they conclude that at the Pd–Au interfaces, the Au layers are slightly compressed and the Pd layers slightly expanded. Their calculations predict that above $f = 2\%$, Pd could reveal ferromagnetic behavior. Evaporating a few ML Pd on Ag(100) the misfit could even be enhanced to 5%.

We have evaporated Pd(100) films on atomically clean and flat Ag(100) substrates. From first ECS experiments at uncoated surfaces of 2–100 ML thick Pd films, we find nonzero in–plane ESP at room temperature, which clearly establishes the existence of long–range ferromagentic order at these surfaces [IX.13]. Further, we find that the ESP is very sensitive to hydrogen adsorption at the Pd(100) surfaces. Desorption of the hydrogen restores the former ESP.

IX.3.5 Surface Magnetism of $Fe_{80}B_{20}$ and Ni(110)

In our low–energy ECS experiments, we use small–angle reflection of 25 keV protons at surface of amorphous ferromagnet $Fe_{80}B_{20}$. At room temperature, we find a short–range ESP of 55% [IX.14]. This value is very close to the spin polarization of the electrons emitted near photo–threshold in photo–emission experiments [IX.36] at surfaces of $Fe_{83}B_{17}$.

Using SPEES at surfaces of $Fe_{80}B_{20}$, we find angle– and energy–integrated P values between $+15\%$ and $+20\%$ which show that these surfaces are ferromagnetic [IX.14]. We used an angle of incidence of $3°$. The obtained ESP data compare favorably with the value of the total average magnetization $(+21\%)$ [IX.37]. For such large angles of incidence, the incident ions can penetrate into the solid and induce the emission of electrons also from layers located deeper than the topmost surface layer.

To obtain a deeper insight into the specific physical processes underlying SPEES, we performed SPEES experiments at (110)–surfaces of Ni picture–frame single crystals. From our ECS experiments [IX.5] and from other spin–sensitive experiments at Ni(100) [IX.5], it is known that electrons originating from energy levels near (< 0.5 eV) the Fermi energy possess a predominant *minority* spin orientation (negative ESP) of about -96%, whereas electrons from energy levels below this energy range overwhelmingly possess a *majority* spin orientation (positive ESP). This information can be profitably used to identify and unravel the nature of the various physical processes involved in the electron emission in SPEES. Electrons originating in k–space from energy levels located 0.5 eV near the Fermi level, should contribute to the *negative* part of the $P(E)$ curve. On the other hand, electrons originating from levels located at least 0.5 eV below the

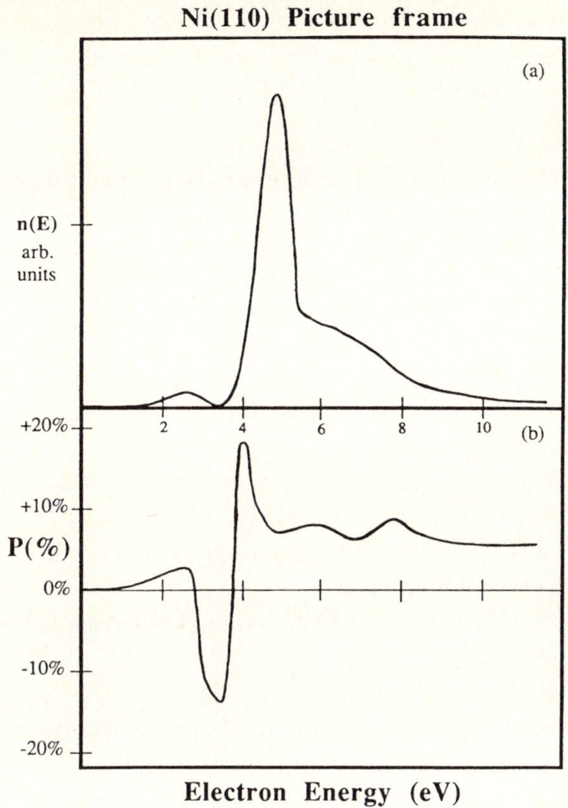

Figure IX.3. Energy distribution $n(E)$ (Fig. IX.3a) and spin polarization $P(E)$ (Fig. IX.3b) of electrons emitted during grazing–angle surface reflection of 25 keV protons at magnetized surfaces of Ni(110) picture frame crystals as function of the electron energy.

Fermi level would contribute to the *positive* part of the $P(E)$ curve. Reducing, for 25 keV incident protons, the angle of incidence from 3° to 1°, thereby preventing near completely the penetration of the protons into the solid, results in a characteristic change in the energy distribution $n(E)$ of the emitted electrons.

In Fig. IX.3, the energy distribution $n(E)$ and the electron spin polarization $P(E)$ is given as function of the energy of electrons emitted along the surface normal of Ni(110) at 300 K. The energy distribution of the ejected electrons, which is nearly independent of the energy of the primary ions, is characterized by a low–intensity peak at around 2.5 eV, followed by a maximum at around 4.5 eV and a decrease of $n(E)$ with increasing energy reaching zero at around 12 eV. The corresponding spin distribution $P(E)$ of the emitted electrons shows several pronounced, characteristic peaks.

We note that, at present, a quantitative theory of SPEES does not exist. With such a theory, one could directly link the measured electron energy and spin distribution to the spin–polarized electron densities of state at the topmost surface layer of ferromagnetic material.

This work was supported by the National Science Foundation, the Robert A. Welch Foundation and the Texas Higher Education Coordination Board.

References

IX.1 For a review, see: M. E. Fisher, *Proc. of the Robert A. Welch Foundation Conferences on Chemical Research*, XXIII, Houston, Texas, Nov. 12–14, 1979, ed. by W. O. Milligan (Houston, 1980).

IX.2 H. W. Diehl, in *Phase Transitions and Critical Phenomena*, edited by C. Domb and J. L. Lebowitz (Academic, London, 1986), Vol. 10.

IX.3 *Modulated Structures*–1979, AIP Conference Proceedings, Vol. 53, edited by H. C. Wolfe (AIP, New York, 1979).

IX.4 *Phase Transitions and Critical Phenomena*, edited by C. Domb and J. L. Lebowitz (Academic, London, 1983).

IX.5 C. Rau, J. Magn. Magn. Mat. **30** 141 (1982).

IX.6 C. Rau and G. Eckl, in *Nuclear Physics Methods in Materials Research*, edited by K. Bethge et al. (Vieweg, Braunschweig, 1980) p. 360.

IX.7 C. Rau, C. Schneider, G. Xing, and K. Jamison, Phys. Rev. Lett. **57**, 3221 (1986).

IX.8 C. Rau and G. Xing, J. Vac. Sci. Technol. A **6**, 579 (1988).

IX.9 C. Rau, C. Jin, and G. Xing, Phys. Rev. Lett. (1989), submitted.

IX.10 C. Rau and C. Jin, J. Phys. (Paris) C **8**, 1627 (1989).

IX.11 H. W. Diehl and E. Eisenriegler, Phys. Rev. B **30**, 300 (1984).

IX.12 C. Rau, G. Xing, and M. Robert, J. Vac. Sci. Technol. A **6**, 579 (1988).

IX.13 C. Rau and G. Xing, to be published.

IX.14 C. Rau and K. Waters, Nucl. Instrum. Methods B **33**, 378 (1988).

IX.15 C. Rau and K. Waters, Nucl. Instrum. Methods B **40**, 127 (1989).

IX.16 J. Schou, Nucl. Instrum. Methods **170**, 317 (1980).

IX.17 C. Rau and S. Eichner, Phys. Rev. Lett. **47**, 939 (1981).

IX.18 J. Unguris, D. T. Pierce, and R. J. Celotta, Rev. Sci. Instrum. **57**, 1319 (1986).

IX.19 C. Rau (1974), unpublished.

IX.20 W. Kuemmerle and U. Gradmann, Solid State Commun. **24**, 33 (1977).

IX.21 W. Kuemmerle and U. Gradmann, Phys. Status Solidi A **45**, 171 (1978).

IX.22 R. Halbauer and U. Gonser, J. Magn. Magn. Mat. **35**, 53 (1983).

IX.23 W. Becker, H. D. Pfannes, and W. Keune, J. Magn. Magn. Mat. **35**, 53 (1983).

IX.24 B. T. Jonker, K. H. Walker, E. Kisker, G. A. Prinz, and C. Carbone, Phys. Rev. Lett. **57**, 142 (1986).

IX.25 B. Heinrich, K. B. Urquhart, A. S. Arrott, J. F. Cochran, K. Myrtle, and S. T. Purcell, Phys. Rev. Lett. **59**, 1756 (1987).

IX.26 J. Balberg and J. S. Helman, Phys. Rev. B **18**, 3030 (1978).

IX.27 C. Rau and S. Eichner, Phys. Rev. B **34**, 6347 (1986).

IX.28 G. Yokoyama, N. Hirashita, T. Oguchi, T. Kambara, K. I. Gondaira, J. Phys. F **11**, 1643 (1981).

IX.29 C. L. Fu, A. J. Freeman, and T. Oguchi, Phys. Rev. Lett. **54**, 2700 (1985); C. L. Fu and A. J. Freeman, J. Magn. Magn. Mat. **54–57**, 777 (1986).

IX.30 J. G. Gay and Roy Richter, Phys. Rev. Lett. **56**, 2728 (1986).

IX.31 J. Kolaczkiewicz and E. Bauer, Surf. Sci. **175**, 487 (1986).

IX.32 C. Rau, C. Jin, C. Liu, Vacuum **39**, 129 (1989).

IX.33 H.–W Diehl, private communication.

IX.34 M. B. Brodsky and A. J. Freeman, Phys. Rev. Lett **45**, 133 (1980).

IX.35 M. B. Brodsky, J. Magn. Magn. Mat. **35**, 99 (1983).

IX.36 R. Allenspach, E. Collan, D. Mauri, M. Landolt, and E. P. Wohlfarth, Phys. Lett. **105**A, 145 (1984).

IX.37 F. E. Luborsky, in *Ferromagnetic Materials*, edited by E. P. Wohlfarth (North Holland, 1980) Vol 1, p. 451.

X. Incommensurable Surface Spin Structure in MnO-Type Antiferromagnets

L.M. Falicov and D.C. Chrzan

Department of Physics, University of California, Berkeley, CA 94720, USA, and Materials and Chemical Sciences Division, Lawrence Berkeley Laboratory, Berkeley, CA 94720, USA

A theory for the phase stability of incommensurable spin structures on the $\{001\}$ surfaces of the rock–salt antiferromagnets is presented. It consists of classical spins and a simple Heisenberg Hamiltonian dependent on three exchange interactions: (a) a surface–only nearest–neighbor exchange; (b) a surface–second–layer nearest–neighbor exchange; and (c) an antiferromagnetic second–nearest–neighbor superexchange throughout the crystal. Incommensurable magnetic surface structures are proven to be ground state for a wide range of the surface exchange parameters.

X.1. Introduction

The Europium monochalcogenides are magnetic semiconductors [X.1]. They display a variety of magnetic behaviors: EuO and EuS are ferromagnets, EuTe is an antiferromagnet, and EuSe is ferromagnetic below 2.8 K, and antiferromagnetic [X.2] between 2.8 K and 4.6 K.

The magnetism in these compounds arises primarily from exchange interactions involving the localized $4f$–shell electrons of the Eu atoms [X.1]. In the rock–salt structure of EuX, where X is O, S, Se, or Te, the Eu atoms are located on a face–centered–cubic lattice. The varied magnetic structures observed in the EuX compounds are a consequence of the competition between dipole–dipole interactions and three exchange processes: (i) the direct overlap of the hybridized Eu $4f$–$5d$ orbitals with the twelve neighboring Eu orbitals (generally ferromagnetic); (ii) the superexchange interaction [X.4] through the valence band formed largely by p orbitals of X (antiferromagnetic); and (iii) a nearest–neighbor exchange through the conduction band (either antiferromagnetic or ferromagnetic depending on the amount of doping). Because of the axial nature of the p orbitals of X, the superexchange mechanism is strongly directed and vanishes for nearest–neighbor Eu atoms [X.6]. The resulting stable magnetic structures consist of (111) ferromagnetically aligned planes, with alternate planes aligned either parallel (ferromagnets), or antiparallel (antiferromagnets) to each other. The antiferromagnetic structure has been observed [X.6] in the transition–metal oxides (NiO, CoO, MnO, and FeO), and in EuTe.

The surfaces of these compounds display anomalous magnetic properties [X.8,X.9]. Techniques which probe the surface magnetic structure either di-

Springer Proceedings in Physics, Vol. 50   **Magnetic Properties of Low-Dimensional Systems II**
Editors: L.M. Falicov · F. Mejía-Lira · J.L. Morán-López © Springer-Verlag Berlin, Heidelberg 1990

rectly (e.g. low–energy electron diffraction (LEED) [X.8,X.10–X.12], spin–polarized low–energy electron diffraction (SPLEED) [X.8]), or indirectly, (e.g. spin–polarized photoemission [X.13–X.18]) have provided valuable experimental results. In the experiment which prompted this research [X.19], Grazhulis and collaborators report the appearance of symmetry–breaking incommensurable surface spin–structures with temperature dependent wavevectors in low–temperature (≈ 10 K) LEED studies of single–crystal EuTe $\{001\}$ surfaces obtained by cleavage under ultrahigh vacuum conditions.

The calculation presented here [X.20,X.21] demonstrates that the stability of the incommensurable magnetic structures on the $\{001\}$ surfaces of EuTe, observed by Grazhulis and coworkers, most likely originates in the competition between relatively large surface nearest–neighbor exchanges and the second–nearest–neighbor superexchange interactions characteristic of the bulk. The calculation, based on a classical Heisenberg Hamiltonian at zero temperature, includes all possible commensurable structures plus one class of incommensurable surface spin arrangements; it yields a complex phase–stability diagram (as a function of surface exchange integrals) with regions of commensurable and incommensurable ground–state–structures.

X.2. Calculations

Three exchange integrals enter the calculation: J, the superexchange between second–nearest neighbors throughout the crystal; K, the net exchange between nearest neighbors on the surface; and L, the net exchange between nearest neighbors where one atom is in the surface layer, and other is in the second layer. Because only the antiferromagnets are considered, J is restricted to be positive, but K and L are allowed to have either sign. Nearest–neighbor exchange in the bulk is neglected and all layers, except the two surface layers, are assumed to have the bulk antiferromagnetic configuration.

The total energy is written

$$E = J \sum_{(ij)} \mathbf{S}_i \cdot \mathbf{S}_j + K \sum_{\langle ij \rangle} \mathbf{S}_i \cdot \mathbf{S}_j + L \sum_{[ij]} \mathbf{S}_i \cdot \mathbf{S}_j , \qquad (X.1)$$

where $\mathbf{S}_i$ is a classical spin of unit magnitude fixed at site i, (ij) designates a second–nearest–neighbor pair, $\langle ij \rangle$ is a nearest–neighbor pair with both spins at the surface, and $[ij]$ is a nearest–neighbor pair with one spin at the surface and one in the second layer; the sums run over an infinite half space.

The two–dimensional unit cell chosen for the calculation contains four atoms from each plane. The cell, with linear dimension b, and its Brillouin zone are shown in Fig. X.1. (The spins are depicted in the chosen bulk configuration.) The points Y and Y′ in the Brillouin zone are not equivalent because the spin domain structure of the bulk introduces a preferred direction on the surface.

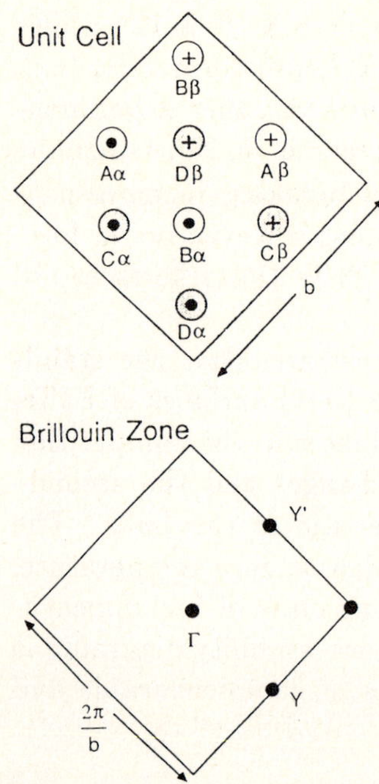

Figure X.1. The unit cell and the Brillouin zone used for the calculation. All spins (indicated in stereographic projections) are in the chosen bulk configuration. The square unit cell has linear dimension b. The first label on each atom refers to each of the four simple–cubic sublattices, and the greek label refers to each of the two face–centered–cubic sub–sublattices. The shaded atoms (labeled C and D) lie in the layer immediately below the surface, the remaining pictured spins (labelled A and B) are in the surface layer. The Γ–point corresponds to all the so–called commensurable structures. The points Y and Y' are not equivalent because of the asymmetry of the bulk spin domain structure.

The Eu face–centered–cubic lattice is divided into four interpenetrating simple–cubic lattices each of which is further divided into two interpenetrating face–centered–cubic lattices. Each simple–cubic sublattice is denoted by a subscript i which runs from A to D. Each face–centered–cubic sub–lattice corresponding to a given simple–cubic sublattice is designated by the subscript μ, which is either α or β.

The trial spin configurations in the two topmost layers have the form of a "frozen", finite–amplitude spin–wave:

$$\mathbf{S}_{i\mu}(\mathbf{R}) = x_{i\mu}\cos(\mathbf{k}\cdot\mathbf{R}+\phi_{i\mu})\hat{\mathbf{x}} + y_{i\mu}\sin(\mathbf{k}\cdot\mathbf{R}+\phi_{i\mu})\hat{\mathbf{y}} + z_{i\mu}\hat{\mathbf{z}} \qquad (X.2)$$

$$x_{i\mu}^2 = y_{i\mu}^2 = 1 - z_{i\mu}^2 ,$$

where $\hat{\mathbf{z}}$ is a unit vector in the direction of the bulk spin quantization, $\mathbf{R}$ refers to the position of the unit cell, and $\mathbf{k}$ lies in the Brillouin zone of Fig. X.2. States with $\mathbf{k} = 0$ are referred to as commensurable, and states with $\mathbf{k} \neq 0$ are called incommensurable. The spins of (X.2) have magnitude unity and the energy given by (X.1)–(X.2) is easily summed to obtain a closed expression for the energy per unit cell for all $\mathbf{k}$, including those at the zone edge.

All spins not in the top two layers are kept fixed:

$$x_{i\mu} = y_{i\mu} = 0 , \qquad (X.3a)$$

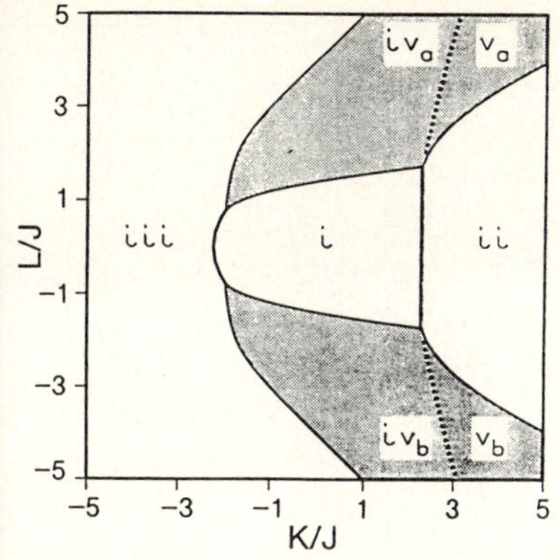

Figure X.2. The phase–stability diagram for all examined structures. Regions (i), (ii), and (iii) are commensurable structures. The shaded regions are incommensurable structures. The incommensurable structures all, as found, have a single **k**–vector. The regions (iv) have an extra degeneracy not present in regions (v).

and

$$z_{i\alpha} = 1, \qquad z_{i\beta} = -1. \tag{X.3b}$$

The total energy (X.1)–(X.2), for given values of (K/J) and (L/J) in the range $-5 \leq (K/J) \leq 5$ and $-5 \leq (L/J) \leq 5$, is minimized with respect to $x_{i\mu}$, $y_{i\mu}$, **k**, and $\phi_{i\mu}$.

X.3. Results and Discussion

The minimum–energy phase–stability diagram is shown in Fig. X.2. It contains commensurable [$\mathbf{k} = 0$ in (X.2), unshaded regions in the figure] as well as incommensurable spin structures [$\mathbf{k} \neq 0$ in (X.2), shaded regions in the figure]. Because all commensurable structures have been included and explicitly calculated, the ground states in the shaded regions are guaranteed to be incommensurable. Since the trial state (X.2) does *not* include *all possible incommensurable structures*, the true *incommensurable* ground states may be different from the ones reported here.

The structures labeled (i), (ii), and (iii) are all commensurable, i.e. the unit cell of the surface structure shown in Fig. X.1, with a given arrangement of the eight spins, repeats itself periodically. The incommensurable structures, labeled (iv) and (v), are of two types. The stable structures in regions (iv) are the finite–amplitude "frozen" spin–waves, whose z–components are reminiscent of the bulk antiferromagnetic state, i.e. a state where the surface spins are arranged in diagonal ferromagnetic stripes with an alternating antiferromagnetic arrangement. The structure appearing in regions (v) are also "frozen" spin–waves, but their z–components are suggestive of a cross between the bulk–antiferromagnetic state and a state with the surface spins arranged in a perfect nearest–neighbor square antiferromagnet (NNSA), with the spins all pointing in the $\pm z$ direction. The subscripts a and b refer to the manner in which the second layer spins align themselves with the surface layer, i.e. generally antiparallel or parallel, respectively. Fig. X.3 is an example of a commensurable

Figure X.3. A stereographic projection of the surface spins for a structure typical of region (ii) of Figure X.2. The dots denote spins pointing up, and the crosses spins pointing down. The tendency toward a surface nearest–neighbor square antiferromagnet in this state is clearly evident.

Figure X.4. A stereographic projection of the surface spins for a structure typical of region (iv_a). The dots denote spins pointing up, and the crosses spins pointing down. The arrow indicates the direction of **k** for this state. This surface state has a character similar to the bulk configuration, and is degenerate with the state pictured in Fig. X.5.

Figure X.5. A stereographic projection of the surface spins for a structure typical of region (iv_a) corresponding to the same spin parameters as used for Fig. X.4. The dots denote spins pointing up, and the crosses spins pointing down. The arrow indicates the direction of **k** for this state. This surface state has a character similar to the bulk configuration, and is degenerate with the state pictured in Fig. X.4.

Figure X.6. A stereographic projection of the surface spins for a structure typical of region (v_a). The dots denote spins pointing up, and the crosses spins pointing down. The arrow indicates the direction of **k** for this state. A tendency towards a z–oriented nearest–neighbor square antiferromagnet is evident.

surface spin structure in region (ii). Figs. X.4, X.5, and X.6 are examples the incommensurable spin structures corresponding to regions (iv_a), (iv_a) and (v_a) respectively. The structures in Figs. X.4 and X.5 are, in fact, for the same values of the parameters; they have the same energy, even though their $\mathbf{k}$–vectors are orthogonal to each other.

The $\mathbf{k}$–vectors of the minimum–energy incommensurable states lie along either the line from Γ–to–Y or the line from Γ–to–Y′ (Fig. X.1). By symmetry, the minimum–energy states with wavevectors $\pm\mathbf{k}$ are degenerate. The structures in regions (iv) have an additional degeneracy: the minimum–energy state with wavevector on the line from Γ–to–Y is degenerate with the state with wavevector of the same magnitude on the line from Γ–to–Y′. This degeneracy is somewhat surprising given the domain asymmetry of the bulk configuration, but it can be easily understood [X.21] based on the fact that only nearest–neighbor spin interactions influence the incommensurable structure.

The value of $\mathbf{k}$ for the minimum–energy state is very sensitive to changes in the surface exchange integrals. Extreme sensitivity occurs in the region of parameter space near the (i)–(iii)–(iv) triple–phase–points and more generally near all the commensurable–incommensurable phase boundaries. The exchange parameters describing the surface of EuTe may be near the (i)-(iii)-(iv_b) triple–phase–point (i.e. antiferromagnetic second–nearest–neighbor exchange and ferromagnetic nearest–neighbor exchanges [X.3]), and hence the small changes in the nearest–neighbor surface exchange expected to arise from temperature variations could generate large, experimentally observable shifts in $\mathbf{k}$.

A notable feature of the results presented here is that the nearest–neighbor coupling L between the surface and second layers is necessary for the stability of the incommensurable "frozen" spin waves. The surface–only nearest–neighbor exchange K, however, is not required for their stability.

X.4. Conclusions

The phase–stability diagram of the simple classical Heisenberg Hamiltonian (X.1), found with trial states of the form (X.2), is remarkably complex. It shows entire regions of parameter space in which incommensurable spin structures are the stable ground state. Since *all* commensurable structures are included in this model, the incommensurable regions of the phase–stability diagram (Fig. X.3) are certain to have incommensurable ground states, which may be the "frozen" spin waves of equation (X.2), or more complex *incommensurable* structures. These incommensurable surface structures are not stabilized by Fermi–surface–type effects, incommensurable or mean–field potentials, but rather are the result of competing nearest– and second–nearest–neighbor interactions. Nearest–neighbor coupling between the first and second layers seems to be necessary for the stability of the incommensurable structures.

It is possible to choose the parameters (K/J) and (L/J) to stabilize the state of any $\mathbf{k}$–vector along the Γ–to–Y, or the Γ–to–Y' line. In some regions of parameter space, which may also coincide with the parameters corresponding to EuTe, the $\mathbf{k}$–vector of the incommensurable stable state is very sensitive to small changes in (K/J) and (L/J).

Since the LEED patterns of these antiferromagnets are expected to display additional diffraction beams caused by magnetic structure at the surface, the magnetic structure factors for several interesting cases were calculated. The results, published elsewhere [X.21], reveal that the LEED pattern should be very sensitive to changes in surface exchange integrals. This sensitivity, expected in both location and intensity of the diffraction beams, should be easily observed.

This research was supported, at the Lawrence Berkeley Laboratory, by the Director, Office of Energy Research, Office of Basic Energy Sciences, Materials Science Division, U. S. Department of Energy, under Contract No. DE-AC03-76SF00098.

References

X.1 *New Developments in Semiconductors*, edited by P. R. Wallace, R. Harris, and M. J. Zuckermann (Noordhoff, Leyden, 1973).

X.2 R. Ritter, L. Jansen, and E. Lombardi, Phys. Rev. B **8**, 2139 (1973).

X.3 T. Janssen, Phys. kondens. Mater. **15**, 142 (1972).

X.4 P. W. Anderson, Phys. Rev. **115**, 2 (1959).

X.5 M. A. Ruderman and C. Kittel, Phys. Rev. **96**, 99 (1954).

X.6 B. Koiller and L. M. Falicov, J. Phys. C **8**, 695 (1975).

X.7 C. Kittel, *Introduction to Solid State Physics*, 6th ed. (Wiley, New York, 1986), p. 444.

X.8 *Magnetic Properties of Low–dimensional Systems*, edited by L. M. Falicov and J. L. Morán–López (Springer–Verlag, Heidelberg, 1986).

X.9 D. Castiel, Surf. Sci. **60**, 24 (1976).

X.10 V. A. Grazhulis, A. M. Ionov, and V. F. Kuleshov, Phys. Chem. Mech. Surfaces **4**, 503 (1986).

X.11 R. E. De Wames, Phys. Status Solidi **39**, 437 (1970).

X.12 L. M. Falicov and R. E. De Wames, Phys. Status Solidi **39**, 445 (1970).

X.13 H. C. Siegmann, Phys. Rev. **17**, 37 (1975).

X.14 R. H. Victora and L. M. Falicov, Phys. Rev. B **31**, 7335 (1985).

X.15 L. E. Klebanoff, R. H. Victora, L. M. Falicov, and D. A. Shirley, Phys. Rev. B **32**, 1997 (1985).

X.16 K. Sattler and H. C. Siegmann, Phys. Rev. Lett. **29**, 1565 (1972).

X.17 M. Campagna, D. T. Pierce, K. Sattler, and H. C. Siegmann, Helv. Phys. Acta. **47**, 27 (1974).

X.18 S. F. Alvarado, E. Kisker, and M. Campagna, in *Magnetic Properties of Low-dimensional Systems*, edited by L. M. Falicov and J. L. Morán-López (Springer-Verlag, Heidelberg, 1986) p. 52.

X.19 V. A. Grazhulis, private communication.

X.20 D. C. Chrzan and L. M. Falicov, Phys. Rev. Lett. **61**, 1509 (1988).

X.21 D. C. Chrzan and L. M. Falicov, Phys. Rev. B **39**, 3159 (1989).

XI. Magnetic Properties of Anisotropic Thin Films

J.M. Sánchez[1] *and J.L. Morán-López*[2]

[1]Columbia University, New York, NY 10027, USA
[2]Instituto de Fisica, Universidad Autónoma de San Luis Potosi,
  78000 San Luis Potosi, S.L.P., México

The magnetic properties of Ising ferromagnetic thin films are studied as function of film thickness and the inter– and intralayer interactions. The magnetic transition, and the total and surface zero–field susceptibilities are calculated for the (0001) hexagonal close–packed structure using the tetrahedron–octahedron approximation of the Cluster Variation Method. The calculations are carried out assuming nearest–neighbor coupling constants. In addition, we assume that i) the coupling constants J_s at the two (0001) surface layers differ in general from the bulk coupling constants and ii) anisotropic effects are included by allowing for different coupling constants parallel (J_1) and perpendicular (J_2) to the surface planes. Special attention is given to finite size effects on the T *vs.* J_s/J_1 phase diagrams, which display the ordinary and extraordinary transitions occurring for enhanced surface coupling constants. The results of the model are discussed in the light of recent experiments in Gd–(0001) ultra–thin films.

XI.1. Introduction

The properties of finite and semi–infinite magnetic films show a variety of new phenomena associated with the reduced dimensionality of the system and the sudden change in environment of the surface atoms. Theoretically, it is now well known that in semi–infinite systems the surface Curie temperature may exceed that of the bulk for values of the surface magnetic coupling J_s larger than a critical value J_{sc} which, in turn, depends sensitively on the surface geometry, bulk structure and range of interactions. This behavior, generally referred to as surface enhanced magnetism, has been the subject of several theoretical studies in simple–cubic and face–centered–cubic structures for a variety of high symmetry surface planes. At present, complete phase diagrams displaying the phenomenon of surface enhanced magnetism in thin films and semi–infinite systems have been determined using Monte Carlo [XI.1,XI.2], renormalization group [XI.3–XI.5], the molecular–field approximation [XI.6–XI.8], and the cluster variation method [XI.9,XI.10]. In general, the predictions of the theoretical models agree with recent experimental results in Ni(100) [XI.11] and Fe on a Au(100) substrate [XI.12], for which no surface enhanced magnetism is observed, and in thin films of Cr [XI.13] and Gd [XI.14–XI.16] for which surface Curie temperatures larger than the bulk have been reported.

Springer Proceedings in Physics, Vol. 50   **Magnetic Properties of Low-Dimensional Systems II**
Editors: L.M. Falicov · F. Mejía-Lira · J.L. Morán-López © Springer-Verlag Berlin, Heidelberg 1990

Although the comparison between simple statistical models and experiment, particularly for ultra–thin films, is hampered by the expected effect of thickness on the magnetic coupling constants and substrate–film interactions, these simple models can be effectively used in order to infer general trends. However, most theoretical studies are limited to cubic crystal structures and, thus, little is known about the effect of complex surface geometries and anisotropy on the thermodynamic properties of these magnetic films. Here we carry out a study of the magnetic properties of ferromagnetic thin films with the hexagonal close–packed structure in the bulk and a (0001) surface. The finite–temperature thermodynamic properties of these ferromagnetic thin films are determined using the tetrahedron–octahedron approximation of the cluster variation method. The choices of crystal structure and surface are motivated by recent experimental results in thick Gd(0001) films using spin–polarized low–energy electron diffraction (SPLEED) [XI.15] and electron capture spectroscopy (ECS) [XI.14], as well as susceptibility measurements in ultra–thin films [XI.17,XI.18]. The model and the statistical mechanics approximation employed are presented in Section 2. The results are presented and discussed in Section 3, and concluding remarks are given in Section 4.

XI.2. The Model

We consider a free (no substrate) spin–1/2 Ising ferromagnetic thin film with N layers and hexagonal close–packed structure in the bulk and (0001) surfaces, characterized by the following Hamiltonian:

$$H = -J_S \sum_{n=1,N} {\sum_{i,j}}' S_i^n S_j^n - J_1 \sum_{n=2}^{N-1} {\sum_{i,j}}' S_i^n S_j^n - J_2 \sum_{n=1}^{N-1} {\sum_{i,j}}' S_i^n S_j^{n+1} \quad (XI.1)$$

where J_S and J_1 are, respectively, the nearest–neighbor ferromagnetic intralayer interactions on the surface and bulk layers and where J_2 is the nearest–neighbor interlayer coupling. Correspondingly, the sums in Eq. (XI.1) are over all nearest–neighbor spins S_i^n ($S_i^n = +1, -1$) at site i and layer n.

The finite–temperature behavior of the Hamiltonian of Eq. (XI.1) was determined using the tetrahedron–octahedron (T–O) approximation of the cluster variation method. In the T–O approximation, the two basic clusters considered are the first nearest neighbor regular tetrahedron and octahedron containing first and second nearest neighbors. The clusters, projected on the (0001) plane of the hcp structure are shown in Fig. XI.1. The octahedron consists of the points 1, 2, 3, 5, 6 and 7, and the tetrahedron is formed by the points 1, 2, 4 and 5. In this figure, the atoms 1, 3 and 6 are located at the surface and the interaction between them is J_S.

The cluster variation method, originally proposed to investigate bulk Ising ferromagnets [XI.19], has been extended by the authors to study thermodynamic

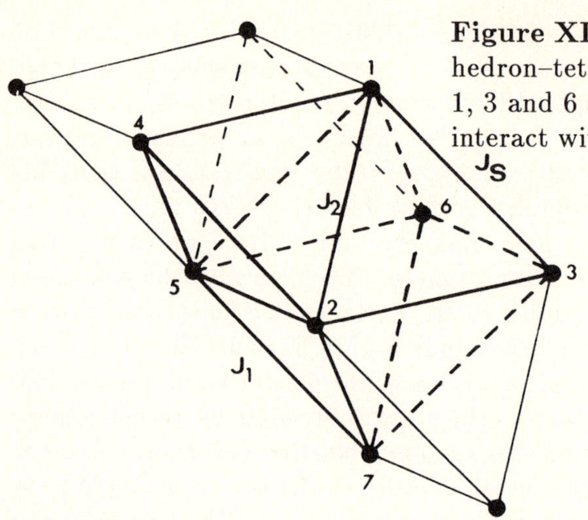

Figure XI.1. Basic cluster used in the octahedron–tetrahedron approximation. The sites 1, 3 and 6 are located at the (0001) plane and interact with a coupling J_s.

properties of surfaces in alloys [XI.20] and in magnetic systems [XI.9–XI.10]. For the cases of bulk three–dimensional spin–1/2 Ising ferromagnets in the *fcc* and isotropic *hcp* structures, the T–O approximation gives a Curie temperature of $kT_C/|J| = 10.0073$, which is in good agreement with high– and low–temperature expansions estimate of $kT_C/|J| = 9.796$; i.e. 2% higher. Similar agreement with exact results is found in the limiting cases of one and two layers. For one layer, the system corresponds to the two–dimensional triangular lattice and the T–O approximation reduces to that of a nearest–neighbor triangular cluster, giving a Curie temperature of $kT_C/|J| = 3.915$ (3.637 exact). For two layers, and taking $J_s = J_1 = 0$, the system reduces to a honeycomb lattice giving a Curie temperature of $kT/|J_2| = 1.6092$ (1.5186 exact).

XI.3. Results

Calculations were carried out in order to investigate the effect of film thickness and anisotropy in the bulk exchange interactions (J_2/J_1) on the temperature vs. J_s/J_1 phase diagram. In Fig. XI.2, we show the resulting phase diagram for an isotropic system $(J_1 = J_2)$ and several film thickness, ranging from a monolayer to 100 layers. For relatively thick films (e.g. N=100), the results are essentially indistinguishable from the semi–infinite case and thus the phase diagram displays an ordinary transition at the bulk Curie temperature T_b, a multicritical point (special transition) taking place for an enhanced surface exchange of $J_{sc}/J_1 = 1.715$ and, for larger values of J_s, a surface transition taking place at higher temperatures than the bulk. Also indicated in Fig. XI.2 is the extraordinary transition at T_b in which the bulk undergoes a para–to–ferromagnetic transition in the presence of a ferromagnetic surface region. We point out that in finite films the extraordinary transition does not actually takes

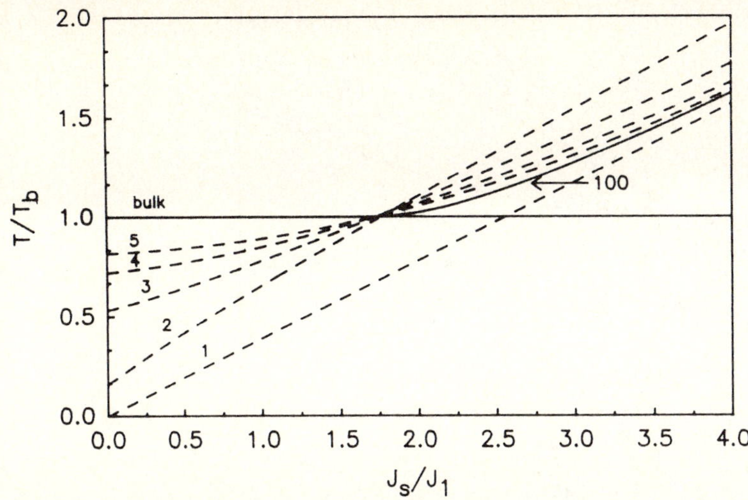

Figure XI.2. Phase diagram in the temperature (T/T_b) *vs.* surface exchange interaction (J_s/J_1) parameter space for the isotropic case. The numbers correspond to the film thickness.

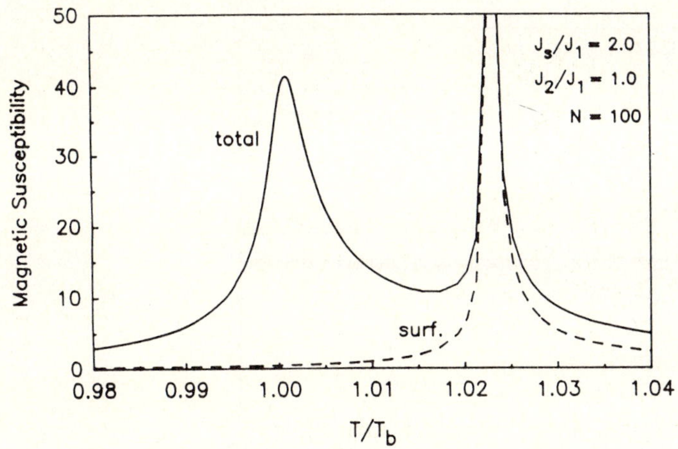

Figure XI.3. The temperature dependence of the zero–field surface (dashed line) and total (continuous line) magnetic susceptibilities for an isotropic 100–layer film. The surface interaction was taken as $J_s = 2J_1$.

place although, for large values of N, the zero–field magnetic susceptibility displays a sharp maximum a that temperature (see Fig. XI.3).

As seen in Fig. XI.2, for small values of J_s the Curie temperature increases with film thickness whereas for large values of J_s the opposite is true. The values of surface exchange at which the crossover point takes place (i.e. when Curie temperature for N layers equals that of N+1 layers), rapidly approaches the critical J_{sc} corresponding to the special transition of the semi–infinite system.

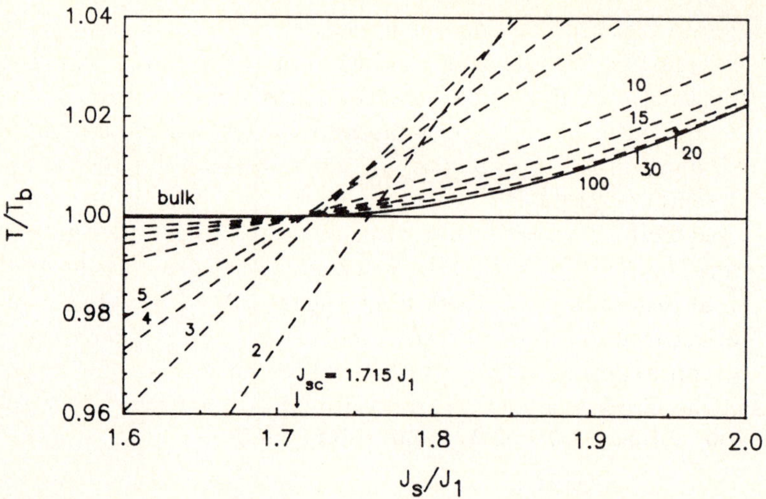

Figure XI.4. Detail of the phase diagram displayed in Fig. XI.2 near the surface critical interaction $J_{sc} = 1.715 J_1$. The numbers correspond to the film thickness.

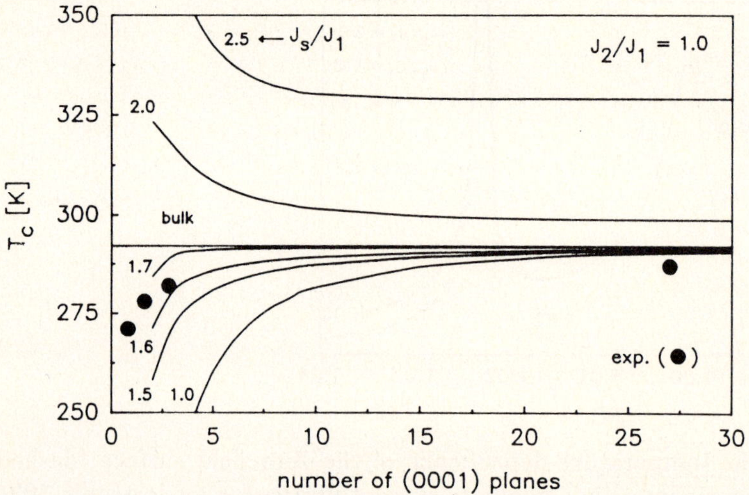

Figure XI.5. Dependence of the transition temperature on the film thickness for various surface– to–bulk–interaction ratios in the isotropic case $J_2 = J_1$. The dots correspond to experimental values of Gd films [XI.17,XI.18].

This behavior is clearly seen in the enlargement (near the special transition) shown in Fig. XI.4.

The variation of the Curie temperature with film thickness for several values of the surface exchange interaction is shown in Fig. XI.5. Also included in Fig. XI.5 are experimental values for thin Gd(0001) films [XI.17,XI.18].

In the light of the simple theoretical model presented here, the experimental results show several interesting features which point towards a complex behavior of the exchange interactions with film thickness and, possibly, anisotropy. In the first place, the experimental results for 0.8 monolayer suggest a significant enhancement in the exchange interactions relative to the bulk value. Furthermore, the change of Curie temperature with N appears to be considerably less pronounced that predicted from size effects only with, apparently, an experimental transition temperature for N=27 appreciably lower than that of the bulk. Finally, the experimental results for very thick films suggest $|J_s| > |J_{sc}|$, as can be inferred from the fact that the surface transition temperature is higher than that of the bulk by approximately 20 K. Within the simple model presented here, the increase in the Curie temperature with film thickness cannot be easily reconciled with the observed surface transition temperature in semi–infinite systems.

We have also carried out calculations for several values of the ratio J_2/J_1. The most significant effect of increasing (decreasing) the anisotropy ratio is to monotonically increase (decrease) the critical value J_{sc} of the surface exchange interaction. The main features displayed by the isotropic phase diagrams of Fig. XI.2 and XI.4 are also present in the anisotropic case and, in particular, a crossover point in the Curie temperature is also found to occur very near J_{sc}. The results of these calculations are summarized in Fig. XI.6 which shows the variation of Curie temperature with film thickness for an anisotropy ratio $J_2/J_1=2$ and several values of J_s/J_1. As seen from the figure, the outstanding discrepancies between the experimental results and the predictions of the

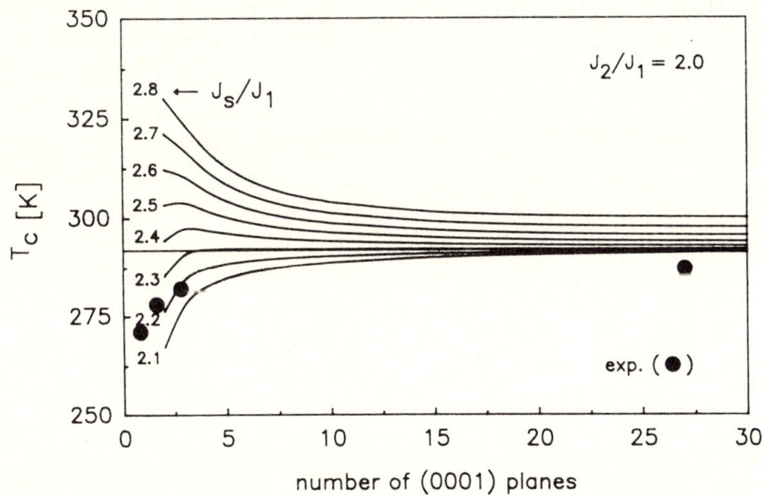

Figure XI.6. Dependence of the transition temperature on the film thickness for various surface– to–bulk–interaction ratios in the anisotropic case $J_2 = 2J_1$. The dots correspond to experimental values of Gd films [XI.17,XI.18].

model are not lifted by the consideration of anisotropy in the bulk exchange interactions.

XI.4. Conclusions

The tetrahedron–octahedron approximation of the cluster variation method was applied to study the phase diagram of hexagonal close–packed thin films with a (0001) surface as a function of film thickness and anisotropy of the bulk exchange interactions. The method offers the distinct advantage of being very accurate and requiring a relatively small computational effort. As such, it allows for the systematic study of complex structure, surface geometries and model Hamiltonians. Our results show that size effects give rise to an increase in the Curie temperature with film thickness for values of surface exchange interactions below J_{sc} whereas the opposite is true for $|J_s| > |J_{sc}|$. This particular behavior fails to reconcile existing experimental data in Gd(0001) films.

This work was supported by the National Science Foundation through Grants No. DMR–8510594 and No. INT–8409776, by Consejo Nacional de Ciencia y Tecnología, Mexico, through Grant No. PCCBBEU–022007, and by Programa Regional de Desarrollo Científico y Tecnológico, OAS.

References

XI.1 K. Binder and D. P. Landau, Phys. Rev. Lett. **52**, 318 (1984).

XI.2 K. Binder, in *Phase Transitions and Critical Phenomena*, Eds. C. Domb and J. L. Lebowitz (Academic Press, 1983), Vol. 8, Chapter 1.

XI.3 H. W. Diehl, S. Dietrich, and E. Eisenriegler, Phys. Rev. B **27**, 2937 (1982).

XI.4 D. F. Sarmento and C. Tsallis, J. Phys. C **18**, 2777 (1985).

XI.5 C. Tsallis, in *Magnetic Properties of Low Dimensional Systems*, Eds. L.M. Falicov and J. L. Morán–López (Springer-Verlag, 1986), SPP Vol. 14, p. 98.

XI.6 D.L. Mills, Phys. Rev. B **3**, 3887 (1971).

XI.7 K. Binder and P. C. Hohenberg, Phys. Rev. B **6**, 3461 (1972).

XI.8 F. Aguilera–Granja and J. L. Morán–López, Phys. Rev. B **31**, 7146 (1985).

XI.9 J. M. Sánchez and J. L. Morán–López, in *Magnetic Properties of Low Dimensional Systems*, Eds. L.M. Falicov and J. L. Morán–López (Springer–Verlag, 1986) SPP Vol. 14, p. 114.

XI.10 J. M. Sánchez and J. L. Morán–López, Phys. Rev. Lett. **58**, 1120 (1987).

XI.11 S. F. Alvarado, M. Campagna, and H. Hopster, Phys. Rev. Lett. **48**, 51 (1982).

XI.12 W. Dürr, M. Taborelli, O. Paul, R. Germar, D. Pescia, and M. Landolt, Phys. Rev. Lett. **62**, 206 (1989).

XI.13 L. E. Klebanoff, S. W. Robey, G. Liu, and D. A. Shirley, in *Magnetic Properties of Low Dimensional Systems*, Eds. L. M. Falicov and J. L. Morán–López (Springer–Verlag, 1986) SPP Vol. 14, p. 75.

XI.14 C. Rau, J. Magn. Magn. Materials, **31–34**, 874 (1983).

XI.15 D. Weller, S. F. Alvarado, W. Gudat, K. Schröder, and M. Campagna, Phys. Rev. Lett. **54**, 1555 (1985).

XI.16 D. Weller and S. F. Alvarado, J. Appl. Phys. **59**, 2908 (1986).

XI.17 M. Farle and K. Baberschke, Phys. Rev. Lett. **58**, 511 (1987).

XI.18 K. Baberschke and M. Farle, Appl. Phys. A **44**, 13 (1987).

XI.19 R. Kikuchi, Phys. Rev. **81**, 998 (1951).

XI.20 J. M. Sánchez and J. L. Morán–López, Surf. Sci. **157**, L297 (1985).

XII. Magnetic Metallic Overlayers on Paramagnetic Substrates

D. Altbir[1], *M. Kiwi*[1], *G. Martínez*[2], *and M.J. Zuckermann*[3]

[1]Facultad de Física, Universidad Católica, Casilla 6177, Santiago 22, Chile
[2]Facultad de Ciencias, Universidad de Chile, Casilla 653, Santiago, Chile
[3]Department of Physics, McGill University, 3600 University Street,
 Montréal, Canada H3A 2T8

Magnetization profiles of thin magnetic overlayers deposited on paramagnetic metal substrates are calculated on the basis of a simple model, which consists of a single site Hubbard Hamiltonian, treated in the Hartree–Fock approximation. The relevant parameters for our model are the degree of band filling, n_f, and the Hubbard intra–atomic Coulomb parameter, U. For weak ferromagnetism (small U) an interesting magnetization *vs.* distance behavior emerges. The feasibility of using a Landau–Ginzburg equation approach is also discussed.

XII.1. Introduction

Over many years magnetic interfaces have attracted the attention of theorists [XII.1–XII.3] and experimentalists [XII.4,XII.5] alike. An important aspect of this work has been the study of composite systems of magnetic metals deposited on paramagnetic substrates. Basically two theoretical approaches have been put forward to treat this problem: the use of Landau–Ginzburg equations (LGE) and microscopic quantum mechanical treatments. In the first method separate LGE for the magnetizations of the magnetic and the paramagnetic metals are written. The magnetizations can then be matched at the interface using boundary conditions analogous to those used by Werthamer [XII.6] and Banarjee [XII.7] for the superconducting proximity effect. This method has already been applied to the problem by one of the authors [XII.8] and suffers from two drawbacks: (i) it can only be used when the coherence length is large compared to the lattice parameter, which confines us to magnetic systems with low Curie or Néel temperatures; and (ii) the boundary conditions of Refs. XII.6 and XII.7 rely on the assumption of continuity of the logarithmic derivative of the order parameter at the interface, which has no *a priori* justification for magnetic systems.

The second method is a microscopic quantum mechanical treatment of the problem formulated by Cox *et al.* [XII.2] and by Mata *et al.* [XII.3]. They considered a system of a magnetic monolayer deposited on a semi–infinite paramagnetic metal. In this contribution we extend these calculations to include a large number of overlayers of magnetic material.

Springer Proceedings in Physics, Vol. 50   **Magnetic Properties of Low-Dimensional Systems II**
Editors: L.M. Falicov · F. Mejía-Lira · J.L. Morán-López © Springer-Verlag Berlin, Heidelberg 1990

XII.2. Model and Solution

The physical system we treat consists of a semi–infinite paramagnetic metal onto which a slab of d layers of a metal, which is magnetic in the bulk, has been deposited. The atoms occupy the sites of a simple cubic structure of lattice constant a. The planes parallel to the surface are labeled by the index l. The values $1 \leq l \leq d$ $(l > d)$ label the magnetic (paramagnetic) layers.

The electronic band structure of the system is characterized by a tight–binding Hamiltonian with a single electron state per atom and spin direction, and by electron hopping between nearest neighbors only. A Hubbard Hamiltonian is used to describe magnetic interactions in both the slab and substrate.

Since the system is invariant under lattice translations parallel to the surface, the mixed Bloch–Wannier representation [XII.9] is used. Treatment of the Hubbard interaction terms in the Hartree–Fock approximation results in the following expression for the effective Hamiltonian: $H_{eff} = \sum_{\vec{K},\sigma} H_{\vec{K},\sigma}$, with

$$H_{\vec{K},\sigma} = \sum_{l=1}^{\infty} \left[\epsilon_{l,\vec{K},\sigma} \, b^{\dagger}_{l,\vec{K},\sigma} \, b_{l,\vec{K},\sigma} + t_{l,l+1} \left(b^{\dagger}_{l,\vec{K},\sigma} \, b_{l+1,\vec{K},\sigma} + h.c. \right) \right] , \qquad (XII.1)$$

where $b^{\dagger}_{l,\vec{K},\sigma}$ $(b_{l,\vec{K},\sigma})$ is the creation (destruction) operator for an electron state in layer l with spin σ and wave vector $\vec{K}$, belonging to the first (two–dimensional) Brillouin Zone (FBZ). The hopping matrix element $t_{l,l+1}$, takes the values t_f, t_{fp} and t_p, in the magnetic slab, at the interface and in the paramagnetic substrate, respectively. In Eq. (XII.1) the following notation was introduced:

$$\epsilon_{l,\vec{K},\sigma} = \begin{cases} \epsilon_0 + 2t_f \left(\cos k_x a + \cos k_y a \right) + U_f \langle n_{l,-\sigma} \rangle & l \leq d \\ 2t_p \left(\cos k_x a + \cos k_y a \right) + U_p \langle n_{l,-\sigma} \rangle & l > d, \end{cases} \qquad (XII.2)$$

where $\langle \ldots \rangle$ denotes the ground state expectation value and k_x, k_y are layer components of $\vec{K}$. The unknown parameters of this Hamiltonian are the atomic level position in the magnetic slab, ϵ_0, the hopping matrix elements in the magnetic material, the interface and the paramagnetic substrate, t_f, t_{fp} and t_p, respectively, and the Hubbard interaction parameters, U_f and U_p. The energies of the atomic levels in the paramagnetic metal are taken to be zero. This way the Hamiltonian has been written as a sum of $2N_{\parallel}$ Hamiltonians $H_{\vec{K},\sigma}$, where $N_{\parallel}$ is the number of atoms per layer parallel to the surface.

Since $\epsilon_{l,\vec{K},\sigma}$ in Eq. (XII.1) depends on the occupancies $\langle n_{l,-\sigma} \rangle$ of each layer, it is necessary to treat the problem self–consistently. This is achieved via an iterative process, which begins with a detailed specification of the physical system. First, numerical values are required for both the thickness of the magnetic slab and the number of paramagnetic layers to be treated exactly (*i.e.* we assume that the perturbation due to the magnetic slab does not extend beyond a certain number of paramagnetic layers). Next, numerical values for the following model parameters must be chosen: the Hubbard intra–atomic parameters, U_f and U_p, the hopping matrix elements, t_f, t_{fp} and t_p, the Fermi level, E_F, and the energy levels ϵ_0 of the magnetic atoms. E_F determines the number of electrons per atom in the bulk paramagnetic metal, while ϵ_0 controls band filling in the slab. The electronic occupation of each layer, for *one* spin direction, completes the specification of the initial configuration of the system. Conventional Green function techniques are used to obtain the relevant physical information.

XII.3. Results

Results of numerical calculations, carried out following the scheme outlined in section 2, are now presented. They are described in terms of the two main parameters: the intra–atomic Coulomb interaction, U_f, of the slab and the parameter, $0 \leq n_f \leq 2$, which describes the filling of the magnetic bands.

Our results fall into two qualitatively different categories, depending on whether the bands of the magnetic slab are exactly half–full or not. The magnetization profiles for the former case are displayed in Figs. XII.1 (see also Fig. XII.3a) and illustrate the behavior of a magnetic slab composed of 35 lay-

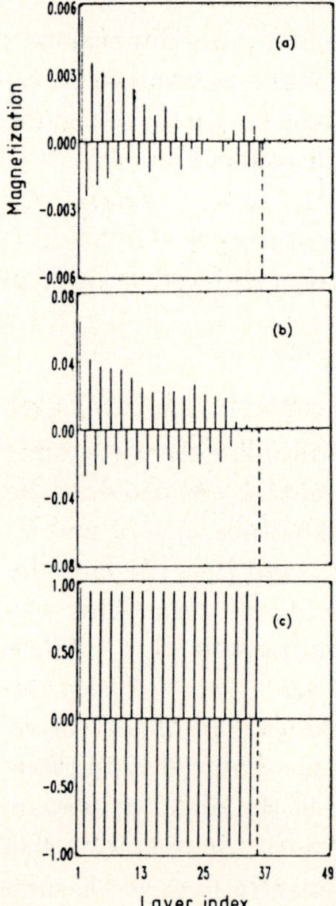

Figure XII.1. Magnetization *vs.* layer index for a slab 35 layers thick. The interface position is indicated by the dashed line. The s–band is exactly half-full ($n_f = 1$). The saturated magnetization equals 1.0. Figs. a, b and c correspond to $U_f = 1.5$, 1.7 and 3.0, respectively.

ers deposited on a semi–infinite paramagnet, for several values of the interaction parameter U_f. For $U_f < 1.7$ the slab exhibits a magnetic order, which is not strictly antiferromagnetic and which fluctuates considerably between neighboring layers (see Figs. XII. 1a,b). For these values of U_f the magnetization per layer is small and effects due to the presence of the surface and/or the interface are important. It is indeed well known [XII.10] that the surface induces variations of the local density of states. When the magnetization is weak, the splitting between the center of the spin–up and spin–down energy bands is quite small and significant changes of the occupation for different spin directions between adjacent layers occur. When the intra–atomic Coulomb correlation reaches $U_f = 2$, the fluctuations relative to the antiferromagnetic order are quite small, as seen in Fig. XII.3a.

Both the magnetization of each layer and the separation between the centers of the minority and majority spin bands increase for larger values of U_f. When this separation becomes larger than the bandwidth as in Fig. XII.1c, variations in the local density of states of the minority spin and majority spin bands do not induce changes in the respective occupation numbers; surface effects become less important and the slab becomes perfectly antiferromagnetic. This result has already been obtained for bulk systems [XII.11] and for two–dimensional clusters [XII.12]. It is directly related to the overlap of the Fermi surface with the magnetic zone boundary which occurs in s–band systems.

When the bands of the slab are no longer half–filled, completely different configurations are obtained. For $n_f = 1.2$ the slab remains paramagnetic up to a critical value of $U_f^c \approx 2$. For $U_f > U_f^c$ the *special ferrimagnetic* configuration shown in Fig. XII.2a is adopted. The transition to this configuration is induced by the creation of magnetic defects in the slab as U_f^c is approached from below. The configuration itself shows a small net magnetic moment with quasi–antiferromagnetic magnetization profiles. In the case of a half–full band, this effect arises from the presence of an interface and/or a free surface, and there is a large increase of the surface magnetization relative to the rest of the slab. As U_f is increased above U_f^c the fluctuations of the magnetization between neighboring planes increase gradually and, for $U_f \approx 3$, the slab exhibits unsaturated antiferromagnetic behavior (Fig. XII.2b). For $U_f \sim 3.6$ the slab undergoes an abrupt transition [XII.12] to the ferromagnetic state of Fig. XII.2c. Any further increase in U_f does not affect the magnetic state of the system, since it has reached saturation.

As n_f increases, the magnetic behavior of the slab can be classified in terms of two possible magnetic structures: paramagnetic, for $U_f \leq 2$ and ferromagnetic for $U_f > 3$. The transition from one to the other is gradual and proceeds via a periodic magnetization of well–defined wavelength. As previously mentioned, the slab becomes ferromagnetic for $U_f > 3$, with small fluctuations between adjacent layers, which decrease with increasing U_f. The stabilization of the ferromagnetic state with increasing U_f can be understood in terms of

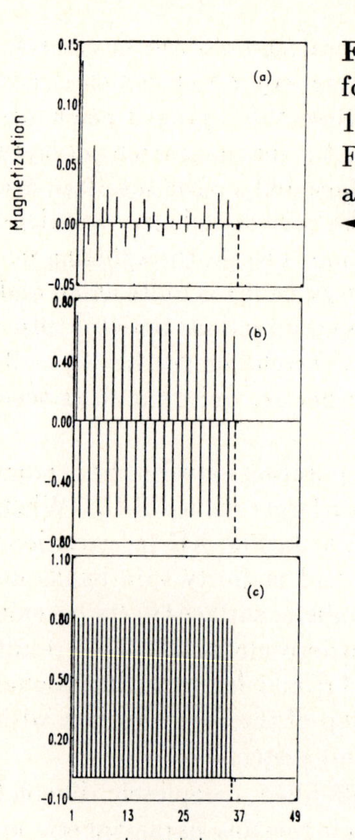

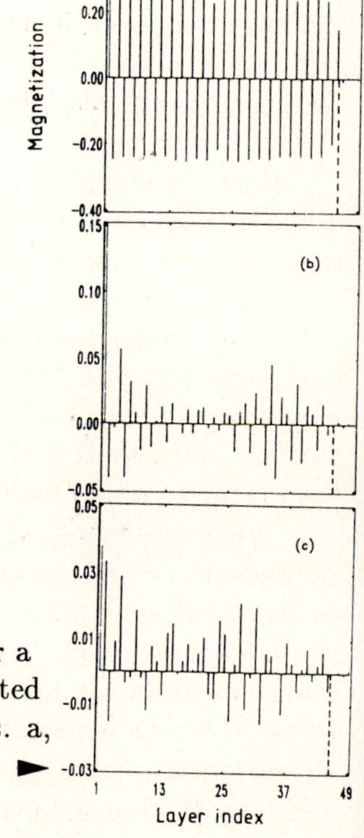

Figure XII.2. Magnetization *vs.* layer index for a slab of 35 layers. The band filling is $n_f = 1.2$; thus saturation magnetization equals 0.8. Figs. a, b and c correspond to $U_f = 2.0$, 3.5 and 3.8, respectively.

◀

Figure XII.3. Magnetization *vs.* layer index for a slab of 45 layers. The interface position is indicated by the dashed line. $n_f = 1.0$, 1.2 and 1.3 in Figs. a, b and c, respectively, and $U_f = 2.0$.

▶

the increase of the separation between the band centers of electronic states with opposite spin directions.

Fig. XII.3 illustrates the effect of varying the band filling parameter n_f on the magnetization profile for a slab composed of 45 layers and characterized by $U_f = 2$. It also shows the magnetization of the first 5 layers of the paramagnetic substrate. Periodic oscillations of the magnetization, which are strongly dependent on the degree of band filling, are clearly noticeable especially in the envelope of the profile. These oscillations are perturbed by the presence of the free surface and the interface, which in turn induce non–commensurate self–consistent potentials.

The dependence of the magnetization profile on the thickness of the slab, for $n_f = 1.2$ and $U_f = 2.0$, is illustrated in Fig. XII.4. For these values the slab

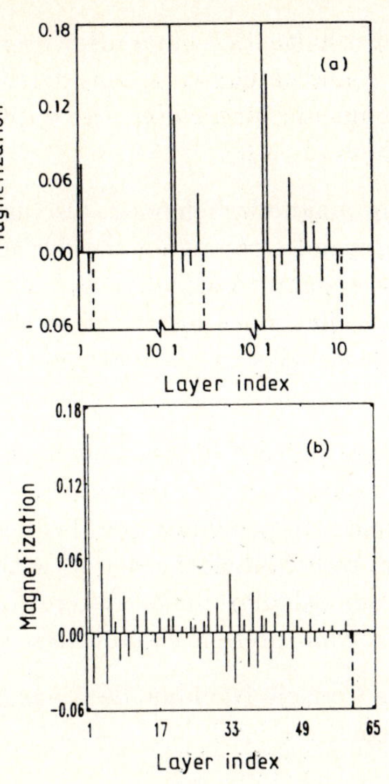

Figure XII.4. Magnetization *vs.* layer index for $n_f = 1.2$, $U_f = 2.0$, and various values of the slab thickness. The interface position is indicated by the dashed line. In Fig. a the magnetization due to slabs of 2, 4 and 10 overlayers is shown on the left, center and right, respectively. In Fig. b the slab is 60 layers wide.

adopts [XII.12] a *special ferrimagnetic* configuration, in which the magnetization direction reverses across almost every layer. Fig. XII.4 illustrates that the modulation pattern, once established, is independent of the number of layers. In our case this requires about 30 layers. (See also Fig. XII.2a). The main consequence of increasing the number of layers is an enhancement of the surface magnetization, while the substrate magnetization remains unchanged. The latter is always weak and decays rapidly as the distance from the interface increases. When the magnetization is saturated all the above conclusions apply equally well.

The above discussion implies that numerical results presented in this section are clearly consistent with the concept of a long–range coherence length (in the sense of the LGE) for small values of U_f and a short–range coherence length for large values of U_f. This can be seen immediately from a comparison of the profiles of Figs. XII.1b and XII.1c. However, the fine details of the various profiles could not have been obtained from Landau–Ginzburg theory.

XII.4. Discussion and Conclusion

The magnetization profiles presented in Sec. XII.3 are consistent with intuitive expectations. In the very weakly magnetic case, *i.e.* for extremely small values of U_f, a collective magnetic ground state is highly unlikely. The only case in which a ground state of this type occurred was for a half–filled band. However, when the more realistic situation of overlapping s–and d–bands is considered, the zone boundary and the Fermi surface never coincide, precluding the onset of magnetic structure.

As U_f increases the slab reveals an interesting magnetic behavior. Oscillatory magnetization profiles are generated, which depend strongly on the degree of magnetic band filling. When U_f becomes large enough to stabilize an anti– or A ferromagnetic structure, surface and interface effects are restricted to the their immediate vicinity. The magnetization profiles in this case are essentially identical to those found in the bulk system.

Periodicities of the order of 10 atomic layers are present in the weak magnetic limit, as observed in Figs. XII.3 and 4. However, the shortness of this wavelength implies that a LGE approach is insufficient to provide a detailed determination of the magnetization profiles. In order to establish the period with greater precision it would be necessary to extend our calculations to systems of at least two hundred atomic overlayers, which is beyond present capabilities.

The support of D. A., M. K. and G. M. by the Consejo Nacional de Ciencia y Tecnología (CONICYT) is gratefully acknowledged.

References

XII.1 M. J. Zuckermann, Solid State Commun. **12**, 745 (1973).

XII.2 B. N. Cox, R. A. Tahir–Kheli and R. J. Elliot, Phys. Rev. B **20**, 2864 (1979).

XII.3 G. J. Mata, E. Pestana and M. Kiwi, Phys. Rev. B **26**, 3841 (1982).

XII.4 G. Bergmann, Phys. Rev. Lett. **41**, 264 (1978).

XII.5 C. Rau, Comments Solid State Phys. **9**, 177 (1980), and references therein.

XII.6 N. R. Werthamer, Phys. Rev. **132**, 2240 (1963).

XII.7 I. Banarjee, Ph. D. Thesis (Northwestern University, 1982) (unpublished).

XII.8 M. J. Zuckermann, Proc. 13 Int. Conf. on Low Temperature Physics, (1972).

XII.9 D. Kalkstein and P. Soven, Surf. Sci. **26**, 85 (1971).

XII.10 M. Kiwi, G. Martínez and R. Ramírez, Acta Metall **84**, 1583 (1988).

XII.11 D. R. Penn, Phys. Rev. **142**, 350 (1966).

XII.12 J. E. Hirsh, Phys. Rev. B **31**, 4403 (1985).

Part 3

Magnetic Superlattices

XIII. Exchange Anisotropy in $Ni_{81}Fe_{19}/Fe_{50}Mn_{50}$ Multilayered Structures: Evidence for Finite-Size Scaling in Ultra-Thin Antiferromagnetic Layers

S.S.P. Parkin[1] *and V.S. Speriosu*[2]

[1]IBM Almaden Research Center, 650 Harry Road, San Jose, CA 95120, USA
[2]IBM Magnetic Recording Institute

The blocking temperature of ultra–thin *antiferromagnetic* (AF) Fe_xMn_{1-x} ($x \simeq 0.5$) layers is determined by their interfacial exchange coupling to a soft ferromagnetic (F) layer of permalloy. The permalloy layer acts as a probe layer which can readily be examined by standard techniques of magnetometry and ferromagnetic resonance. It is observed that the blocking temperature of the AF layer monotonically decreases as the AF layer thickness is reduced, decreasing rapidly for thicknesses below $\simeq 100$ Å. It is shown that this behaviour is consistent with a simple finite–size scaling relationship, assuming the blocking temperature is close to the Néel temperature.

With the development in recent years of increasingly sophisticated techniques to prepare thin films, the physical properties of ultra–thin ferromagnetic films, just a few monolayers thick, have been studied in some detail [XIII.1]. A variety of phenomenological models have predicted drastic changes in magnetic properties of such systems as their physical size is reduced. Perhaps the most basic property of a magnetic material is the temperature of the phase transition below which the material becomes magnetically ordered. For ferromagnetic materials there is a large net magnetic moment. This means that the magnetic properties of thin films with layer thicknesses even as small as one atomic layer can be measured using a variety of experimental techniques including Mössbauer spectroscopy [XIII.1], squid and Kerr magnetometry [XIII.2,XIII.3], anomalous Hall effect [XIII.4], ferromagnetic resonance [XIII.5], and various electron spectroscopies [XIII.6]. For the ferromagnetic first row transition metals, whilst there is some controversy as to the exact thickness dependence of their Curie temperatures, there is widespread agreement that the Curie temperatures of these systems rapidly approach the corresponding bulk values for films just a few monolayers thick [XIII.7].

In contrast to numerous studies of the properties of *ferromagnetic* (F) films there has been very little experimental work on thin *antiferromagnetic* (AF) layers. The absence of a net magnetic moment in AF films precludes using most of the measurement techniques mentioned above or substantially reduces their sensitivity. For example antiferromagnetic resonance, which in any case is useful only in single–crystal films, has been used to study AF thin films,

Springer Proceedings in Physics, Vol. 50   **Magnetic Properties of Low-Dimensional Systems II**
Editors: L.M. Falicov · F. Mejía-Lira · J.L. Morán-López © Springer-Verlag Berlin, Heidelberg 1990

but the sensitivity is such that the thinnest films that have been investigated are more than 1000 Å thick, well above the thickness where important changes in magnetic behaviour are expected [XIII.8]. In the work presented here the properties of much thinner AF layers are examined by taking advantage of a phenomenon first discovered more than 30 years ago in oxidized Co particles and subsequently extensively studied in a number of thin–film coupled systems [XIII.9–XIII.16]. This phenomenon, often referred to as exchange anisotropy, arises from the interfacial magnetic exchange coupling between an AF layer and a F layer which, under certain conditions causes a unidirectional anisotropy in the F layer. Consequently some properties of the AF layer can be indirectly determined from those of the F probe layer which can more easily be measured.

Multi–layered film structures were prepared in a high vacuum DC magnetron sputtering system with a base pressure $\simeq 1 \times 10^{-9}$ torr. A wide range of film structures prepared from ferromagnetic Ni–Fe and antiferromagnetic Fe–Mn alloys were used in these experiments. The structures were of the general form,
$$\mathrm{Si}(111)/\mathrm{Cu}(100\text{Å})/r \times [\mathrm{Ni}_{81}\mathrm{Fe}_{19}(t_{\mathrm{F}})/\mathrm{Fe}_x\mathrm{Mn}_{1-x}(t_{\mathrm{AF}})]/\mathrm{Ni}_{81}\mathrm{Fe}_{19}(t_{\mathrm{F}})/\mathrm{Cu}(100\text{Å}),$$
where $x \simeq 0.5$ and r is the number of times the sequence enclosed within the square brackets is repeated and t_{F} and t_{AF} are the thicknesses of the ferromagnetic and antiferromagnetic layers respectively. Values of r ranged from 1 up to 50. Deposition was carried out at an argon pressure of 3.2×10^{-3} torr at a nominal rate of 2 Å/sec. In one pump–down of the sputtering system up to 20 arbitrarily complex film structures sputtered from up to four different source materials could be prepared via the computer control of shutters and substrate platform position. Deposition was carried out at room temperature with an aligning SmCo magnet placed beneath each substrate to establish a single macroscopic unidirectional anisotropy. For these experiments the films were deposited onto one–inch diameter Si(111) wafers which were chemically etched just prior to placement in the vacuum chamber. The magnetic film structures were sandwiched between underlayers and protective capping overlayers of Cu. The exchange bias field was found to be sensitive to the thickness of the Cu underlayer and was decreased in its absence consistent with earlier work [XIII.16]. The composition of the magnetic layers was determined from SIMS, X–ray fluorescence and EDX studies and was close to that of the composition of the corresponding target material. A vibrating sample magnetometer (VSM) with a sensitivity of $\simeq 2 \times 10^{-6}$ emu was employed to make magnetic hysteresis measurements. Torque curves were obtained with a Toei torque magnetometer with a sensitivity of $\simeq 1 \times 10^{-3}$ dyne–cm. Ferromagnetic resonance (FMR) was carried out at 9.5 and 34.1 GHz using Bruker spectrometers.

The composition of $\mathrm{Fe}_x\mathrm{Mn}_{1-x}$ was chosen to give maximum exchange coupling ($x \simeq 0.5$) following the work of Hempstead et al [XIII.15]. The alloy $\mathrm{Fe}_x\mathrm{Mn}_{1-x}$ takes up two different phases [XIII.17]. The γ phase has the same fcc structure as permalloy and orders antiferromagnetically at about 400–500 K depending on its composition. The α phase has a more complicated structure and is non–magnetic above $\simeq 100$ K. For the structures used in these studies care was taken to ensure that the $Fe_x Mn_{1-x}$ was only of the γ phase. Previous work has shown that films of approximate composition $Fe_{50} Mn_{50}$ when grown onto a conforming layer of Cu take up this structure [XIII.16]. The composition of $Ni_{81} Fe_{19}$ (permalloy) was chosen to give minimum magnetostriction and magnetocrystalline anisotropy [XIII.18]. For simplicity the permalloy and iron–manganese alloys are referred to henceforth as NiFe and FeMn. Sets of related films were grown in one pump–down of the sputtering system to ensure very similar growth conditions. Quartz–crystal microbalances were used to monitor the deposition rates. The absolute film thickness scale was established by making Rutherford backscattering studies (assuming bulk densities) and surface profilometer measurements of 1000 Å single layer calibration films.

The most obvious manifestation of the exchange coupling at the interface between a F and an AF layer is that the magnetic hysteresis loop of the coupled system is shifted along the applied field axis by the exchange anisotropy field, H_B . In the simplest model of this effect, first proposed by Meiklejohn and Bean [XIII.9], an ideal planar F/AF interface is assumed. It then follows that H_B varies inversely with the thickness of the ferromagnetic layer, provided that the exchange anisotropy energy, K_e , is independent of thickness of the F layer. In particular H_B can be described by the relation $H_B = K_e/M_s t_F$. Previous studies have shown good agreement with this model for $t_F \geq 50$ Å [XIII.19,XIII.20]. The behaviour of thinner NiFe layers was investigated by preparing superlattice structures of the form $Cu(100\text{Å})/r \times [FeMn(100\text{Å})/NiFe(t_F)]/FeMn(100\text{Å})/Cu(100\text{Å})$. The value of r was chosen to give the total combined thickness of permalloy closest to 150 Å. In this way it was possible to measure the hysteresis loops of very thin layers of exchange coupled NiFe using conventional magnetometry. Only one hysteresis loop was observed, showing that the properties of NiFe and FeMn sublayers were the same throughout the superlattice. The exchange bias field at 300 K is plotted as a function of $1/t_F$ in Fig. XIII.1a. For t_F above $\simeq 25$ Å the expected relationship $H_B \propto 1/t_F$ is followed with values of $K_e \simeq 7.3 \times 10^{-2}$ erg/cm^{-2} . Surprisingly however K_e $(\equiv M_s H_B t_F)$ drops to zero as the permalloy layer thickness is decreased below $\simeq 25$ Å, even though both VSM and FMR data show that the magnetization of the permalloy layer is close to its bulk value for thicknesses down to $\simeq 10$ Å. These unexpected results suggest that the exchange coupling interaction is not confined to a single layer at the interface but extends over several layers.

The coercive field, H_C , defined as the half–width of the hysteresis loop measured at zero magnetic moment, is also included in Fig. XIII.1a. The dependence of coercive field on t_F is very similar to that of the exchange bias field, showing that the loss mechanisms are dominated by interactions across the interface. Note that as the exchange goes to zero for either very thin or very thick sandwiched permalloy layers, H_C moves towards the value seen in thick, single layer permalloy films ($H_C \simeq 1$ Oe). The maximum coercive field occurs

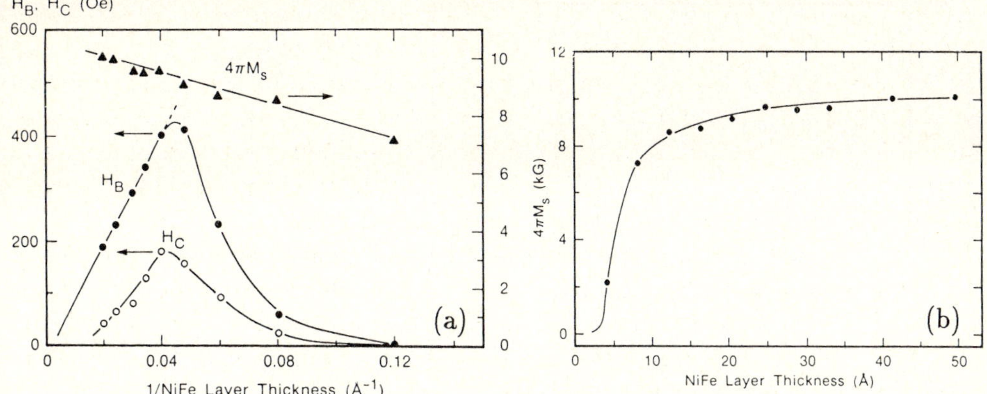

Figure XIII.1. (a) Room temperature exchange bias field, H_B, and coercive field, H_C, as a function of inverse NiFe layer thickness for a series of multi–layered films of structure Si(111)/Cu (100 Å)/$r \times$ [FeMn (100Å)/NiFe(t_F)]/FeMn (100 Å)/Cu(100Å). Also included are room temperature depth–averaged saturation magnetization values, $4\pi M_s$, deduced from FMR. The same $4\pi M_s$ values are plotted *versus* NiFe layer thickness in (b) including points for small thickness.

at a slightly larger NiFe layer thickness than that for the maximum exchange bias field.

The detailed dependence of the saturation magnetization of the permalloy layer on its thickness gives some indication of the quality of the NiFe/FeMn interface on the NiFe side. The depth–averaged saturation magnetization, $4\pi M_s$, was measured by ferromagnetic resonance at 9.5 and 34.1 GHz in the parallel geometry, using the resonance condition $H_r = [(\omega/\gamma)^2 + 2\pi M_s^2]^{1/2} - 2\pi M_s$. Here ω is the resonance frequency and γ is the gyromagnetic ratio. For all samples γ varied little from $1.846 \times 10^7\,\mathrm{Oe}^{-1}\mathrm{sec}^{-1}$, corresponding to $g = 2.10$. As shown in Fig. XIII.1a, at 300 K, $4\pi M_s$ decreases linearly with inverse NiFe layer thickness which can be simply interpreted as consistent with a layer at the NiFe/FeMn interface of reduced magnetization. A second very likely possibility is that the Curie temperature of the layers is reduced as their thickness is reduced, so giving rise to a reduced $4\pi M_s$ at 300 K. Fig. XIII.1b shows $4\pi M_s$ *versus* NiFe layer thickness for NiFe layer thicknesses down to 4Å. These data indicate that the thickness of the NiFe/FeMn interface on the NiFe side of the couple is extremely sharp and no more than $\simeq$ 2–3 Å thick.

Similarly, the thickness of the interface on the FeMn side may be obtained by measuring the FeMn thickness needed to exchange–decouple the two thick NiFe layers of a NiFe/FeMn/NiFe sandwich. To facilitate the measurement, the two NiFe layers were distinguished by alloying one of the NiFe layers with a small amount of Rh, thereby giving it a sufficiently different saturation magnetization, and consequently different ferromagnetic resonance field. The NiFeRh layer was

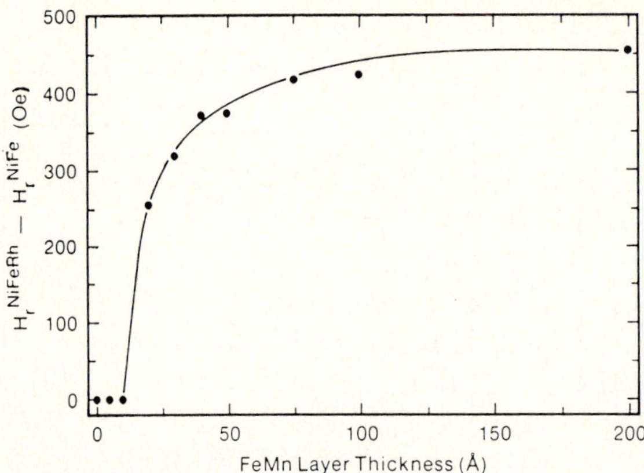

Figure XIII.2.   Difference in room temperature resonance fields at 34.1 GHz in parallel resonance for the NiFeRh and NiFe layers in the sandwich Si(111)/NiFe(100 Å)/FeMn(t_{AF})/NiFeRh(100 Å)/Ta(200 Å) *versus* FeMn layer thickness. The variation of resonance fields with azimuthal angle was averaged out.

of composition $(\mathrm{Ni}_{81}\mathrm{Fe}_{19})_{90}\mathrm{Rh}_{10}$. Fig. XIII.2 shows the difference in resonance fields of the NiFe and NiFeRh layers at 34.1 GHz in the parallel geometry for a series of NiFe(100Å)/FeMn(t_{AF})/NiFeRh(100Å) sandwiches, with t_{AF} varying from 0 to 200 Å. Measurements were made as a function of azimuthal angle in order to average out the effect of exchange anisotropy on the NiFe and NiFeRh layers. For thicknesses of 0, 5 and 10 Å there is only one resonance, consistent with strong coupling of the NiFe and NiFeRh layers. At larger thicknesses the resonance bifurcates, allowing a rough upper estimate of the thickness of one of the NiFe/FeMn interfaces on the FeMn side of $\simeq$ 5–10 Å.

Meiklejohn and Bean gave the following expression for the energy per unit area of an exchange couple consisting of a ferromagnetic and an antiferromagnetic layer [XIII.9,XIII.10]:

$$E = -M_S H t_{\mathrm{F}} \cos(\alpha - \phi_{\mathrm{F}}) - K_e \cos(\phi_{\mathrm{F}} - \phi_{\mathrm{AF}}) + K_{\mathrm{AF}} t_{\mathrm{AF}} \sin^2 \phi_{\mathrm{AF}} \quad (XIII.1)$$

where M_S and H are the saturation magnetization and applied field magnitude, respectively, t_{F} (t_{AF}) is the thickness of the ferromagnet (antiferromagnet), α is the angle of the applied field, ϕ_{F} (ϕ_{AF}) is the angle of the magnetization (the sublattice magnetization for the antiferromagnet), K_e is the interfacial coupling constant, and K_{AF} is the anisotropy (presumed to be crystalline) of the antiferromagnet. For simplicity we have omitted the anisotropy energy of the ferromagnet, since for the NiFe/FeMn couple it is much smaller than K_e. In Eq. (XIII.1) the first term is the energy of the ferromagnet due to the applied field, the second term represents the coupling between the ferro– and antiferromagnet, while the last term is the pinning energy of the antiferromagnet. An important result is that the condition

$$K_{\mathrm{AF}}\, t_{\mathrm{AF}} \geq K_e \qquad\qquad (XIII.2)$$

is required for observation of exchange anisotropy. If Eq. (XIII.2) is not satisfied, the antiferromagnet will follow the motion of the ferromagnet, resulting in unshifted hysteresis loops and uniaxial, rather than unidirectional, torque curves. Consistent with this notion, measurement of hysteresis loops in NiFe/FeMn films as a function of FeMn layer thickness showed [XIII.21,22] a critical AF layer thickness, $t_{cr} \simeq 50$ Å, for the sharp onset of exchange bias field. From the measured value of t_{cr}, the anisotropy of sputtered FeMn, $K_{\mathrm{AF}} \simeq 1.35 \times 10^5$ erg/cm^2, was deduced [XIII.22]. However, this estimate has no meaning since it assumes that K_{AF} is independent of thickness of the AF layer. Detailed torque measurements on similar NiFe/FeMn films show that the existence of a critical thickness in NiFe/FeMn cannot be understood in terms of Eq. (XIII.1) alone, as discussed below.

From Eq. (XIII.1), the torques on the ferromagnetic and antiferromagnetic layers are

$$L_{\mathrm{F}} = -\frac{\delta E}{\delta \phi_{\mathrm{F}}} = M_s\, H t_{\mathrm{F}}\, \sin(\alpha - \phi_{\mathrm{F}}) - K_e \sin(\phi_{\mathrm{F}} - \phi_{\mathrm{AF}}), \qquad (XIII.3a)$$

$$L_{\mathrm{AF}} = -\frac{\delta E}{\delta \phi_{\mathrm{AF}}} = K_e \sin(\phi_{\mathrm{F}} - \phi_{\mathrm{AF}}) - K_{\mathrm{AF}}\, t_{\mathrm{AF}} \sin 2\phi_{\mathrm{AF}}\,. \qquad (XIII.3b)$$

The measured torque is $L_m = M_s\, H t_{\mathrm{F}}\, \sin(\alpha - \phi_{\mathrm{F}})$, which in equilibrium also equals $K_e \sin(\phi_{\mathrm{F}} - \phi_{\mathrm{AF}})$. For large H, when $\alpha \simeq \phi_{\mathrm{F}}$, the dominant term in the measured torque curve varies either as $\sin\alpha$ or $\sin 2\alpha$, depending on whether or not condition (XIII.2) is satisfied. The results of a numerical solution of Eqs. (XIII.3a) and (XIII.3b), shown in Fig. XIII.3, are consistent with these considerations. According to the Meiklejohn–Bean model, as t_{AF} increases from zero, the $\sin 2\alpha$ term increases at first linearly with t_{AF}, reaches a peak at the critical thickness, then drops precipitously to small values. On the contrary, the $\sin\alpha$ term is very small for $t_{\mathrm{AF}} \leq t_{cr}$, jumping to a nearly constant value as $t_{\mathrm{AF}} \geq t_{cr}$. Also shown in Fig. XIII.3 are experimental $\sin\alpha$ and $\sin 2\alpha$ terms obtained from torque curves measured in a 400 Oe field and corrected for the background due to the sample holder. The value of $K_{\mathrm{AF}} = 1.40 \times 10^5$ erg/cm^3 was chosen to agree with the measured t_{cr}, while $K_e = 7.30 \times 10^{-2}$ erg/cm^2 was selected for good agreement with the measured $\sin\alpha$ term for thick AF layers. However, note that as long as $K_{\mathrm{AF}}\, t_{\mathrm{AF}} \geq K_e$, the agreement between calculated and measured $\sin\alpha$ terms (for $t_{\mathrm{AF}} \geq t_{cr}$) is insensitive to the size of K_{AF}. For $t_{\mathrm{AF}} \leq t_{cr}$, Fig. XIII.3 shows a large discrepancy between calculated and measured $\sin 2\alpha$ terms. It is this discrepancy that casts serious doubts about the applicability of the Meiklejohn–Bean model to sputtered NiFe/FeMn. The discrepancy cannot be attributed to a different symmetry, such as fourfold

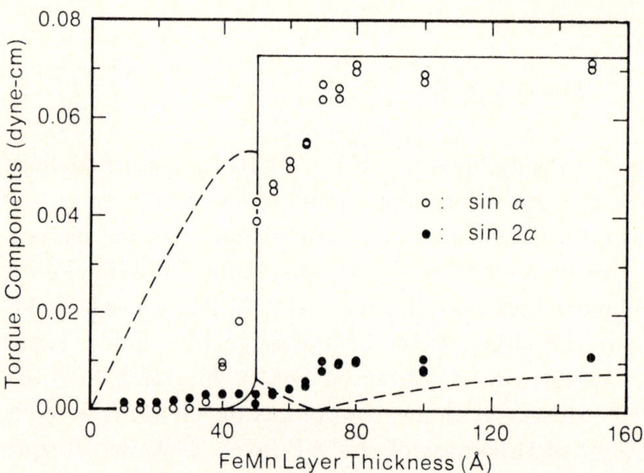

Figure XIII.3. Comparison of the Meiklejohn–Bean model with experiment at room temperature for structures of the form Si(111)/Cu(200 Å)/NiFe(60 Å)/FeMn (t_{AF})/Cu(200Å). Calculated torque terms: $\sin\alpha$ (solid curve), $\sin 2\alpha$ (dashed curve); measured torque terms: $\sin\alpha$ (open circles), $\sin 2\alpha$ (closed circles). No higher order terms were observed in the measured torque curves.

or sixfold, of the anisotropy of FeMn, since measured torque curves do not have appreciable Fourier terms of order higher than $\sin 2\alpha$. In order to reconcile this with Eq. (XIII.1), one must assume that $K_A F$, when measured at room temperature, depends on the thickness of the AF layer. Thus the value of K_{AF} deduced from Eq. (XIII.2) has meaning only at the critical thickness and only at room temperature.

Since the dependence of H_B on AF thickness cannot be accounted for by a low value of the bulk anisotropy energy of the AF layer compared to the exchange coupling energy, an alternative explanation is needed. Perhaps the most straightforward explanation of the loss of exchange anisotropy at 300 K for thin FeMn layers is that the magnetic ordering temperature of this layer, $T_N\,(t_{\mathrm{AF}})$, is reduced from the bulk value. This would mean that for $t_{\mathrm{AF}} \leq 50$ Å the Néel temperature is reduced below $\simeq 300$K. A systematic study of the temperature dependence of exchange bias field as a function of FeMn layer thickness was carried out for two different sets of film structures with slightly different compositions of FeMn. In both cases it was found that indeed the temperature at which the exchange bias field goes to zero, the blocking temperature, T_B, rapidly decreases with decreasing FeMn thickness for $t_{\mathrm{AF}} \leq 100$ Å but for thicker layers saturates and becomes independent of FeMn layer thickness. It must be emphasized that in these experiments it is the blocking temperature of the F/AF couple that is measured and not the magnetic ordering temperature. Whereas the latter is the temperature at which the long–range magnetic order disappears, T_B is the temperature at which condition (XIII.2) above is no longer

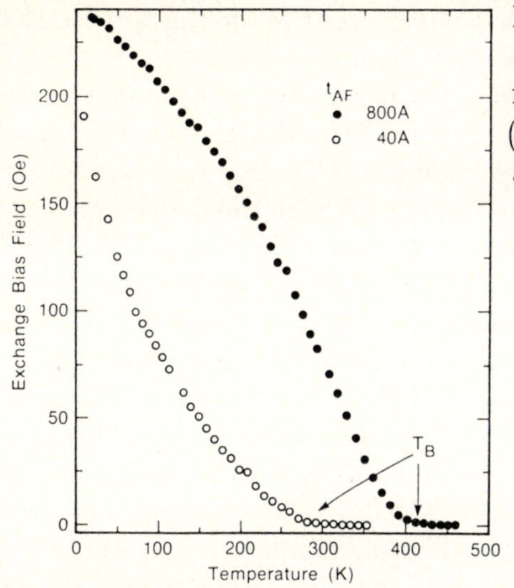

Figure XIII.4. Exchange bias field *versus* temperature for two structures of the form, Si(111)/Cu(100Å)/NiFe(60Å)/FeMn (t_{AF})/Cu(100Å), for values of t_{AF} of 40 and 800 Å.

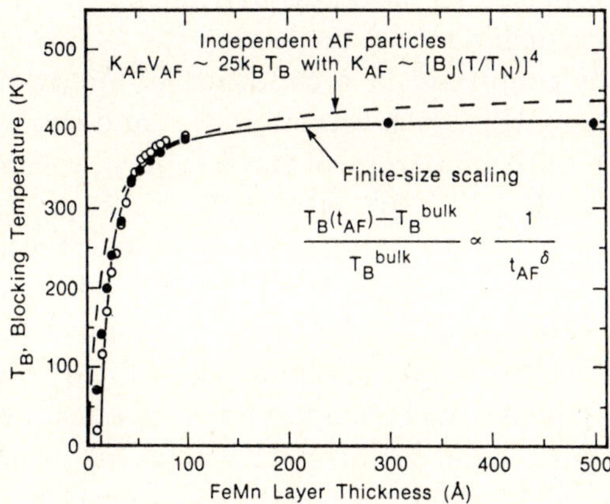

Figure XIII.5. Blocking temperature, T_B, *versus* FeMn layer thickness and comparison with a finite–size scaling model and an independent particle model as discussed in the text. o and • correspond to two sets of films of the same structure but containing FeMn layers with slightly different compositions.

satisfied. Since T_B must be less than but could be very close to T_N (t_{AF}), we assume $T_B \simeq T_N$ (t_{AF}). Typical H_B *versus* temperature data are shown in Fig. XIII.4 and the thickness dependence of the blocking temperature is shown in Fig. XIII.5. The larger data set shown in Fig. XIII.5 includes data for FeMn layer thicknesses extending from 10 to 1000 Å. These data can be very well described by the relation

$$\frac{T_B\left(t_{\mathrm{AF}}\right) - T_B^{\mathrm{bulk}}}{T_B^{\mathrm{bulk}}} \propto \frac{1}{t_{\mathrm{AF}}^{\delta}} \qquad (XIII.4)$$

with $\delta \simeq 1.6 \pm 0.3$. This relation is just that which has been found to describe the thickness dependence of the Curie temperature of thin ferromagnetic layers for films prepared under ideal ultra high vacuum conditions and for those prepared by less sophisticated sputtering techniques [XIII.23,XIII.24]. This type of behaviour can be ascribed to finite–size scaling as the magnetic correlation length becomes bounded by the size of the system. Whilst this behaviour has, to our knowledge, not previously been observed in thin antiferromagnetic films, work on oxidized Ni, Co and permalloy and analogous systems in the sixties and early seventies showed reduced blocking temperatures compared to bulk Néel temperatures [XIII.11–XIII.13]. For these systems the oxide layer thickness was poorly known and of questionable integrity. One model which was developed to account for the properties of such systems assumed the AF oxide coating was comprised of an ensemble of exchange decoupled particles of size close to the superparamagnetic limit [XIII.11–XIII.13]. The blocking temperature of such a system would then be determined by the temperature at which the anisotropy energy of the particles was comparable to the thermal energy in the system. In Fig. XIII.5 we compare the measured thickness dependence of T_B with the closest dependence derived from such a model by making the following assumptions. Firstly, the FeMn layer is comprised of a collection of identical magnetically decoupled particles of volume, V_{AF}, with diameter, d_{AF}, and thickness that increases as t_{AF}. Secondly the bulk anisotropy of the AF, K_{AF}, varies as some power of the Brillouin function, $[B_J\,(T/T_N\,)]^p$, and finally T_B is given by the relation $K_{AF} V_{AF} \simeq 25 k_B\, T_B$ where k_B is Boltzmann's constant. Both d_{AF} and p were varied to give the best agreement with the experimental data which is included in Fig. XIII.5. The values used in the figure correspond to $d_{AF} = 1000$ Å and $p = 4$. As shown in the figure, the independent particle model describes the data more poorly than the much simpler finite–size scaling model. In particular, the independent particle model predicts a much stronger dependence of blocking temperature on t_{AF} for thick AF layers than is observed. For example this model gives $\simeq 10\%$ increase in T_B for t_{AF} increasing from 250 Å to 800 Å whereas no increase is found within experimental error, consistent with the finite–size scaling model. Moreover, since the fitted value of p is much smaller than expected for an antiferromagnet with cubic symmetry [XIII.25], this is clearly not the correct model. Although we cannot rule out the possibility that a more sophisticated superparamagnetic model could account for the data in Fig. XIII.5, the elegance of the finite–size scaling model, and the good agreement with the data are strong points in its favour.

In summary we have demonstrated that by taking advantage of the phenomenon of exchange coupling of a ferromagnetic layer to an antiferromagnetic thin film it is possible to study the properties of ultra–thin AF layers just a few angstrom thick. We have shown that the blocking temperature of an antiferromagnetic FeMn layer (coupled to a permalloy layer) is dramatically decreased when its thickness is reduced below $\simeq 100$ Å. The dependence of the blocking temperature on the thickness of the AF layer can be described by a simple scaling law consistent with finite–size scaling. We propose that exchange coupling provides a general method which can readily be applied to other AF/F layered systems to allow the properties of ultra–thin AF layers to be examined.

We are particularly grateful to K. P. Roche for excellent technical assistance and we are happy to acknowledge V. Deline, G. Gorman, T. C. Huang, D. Miller and R. Savoy for carrying out various structural analyses. We thank Prof. S. H. Charap for drawing our attention to his work.

References

XIII.1 See, for example, *Magnetic Properties of Low–Dimensional Systems*, edited by L. M. Falicov and J. L. Morán–López, Springer Proceedings in Physics **14**, Springer–Verlag, Berlin, 1986.

XIII.2 G. Bayreuther, J. Magn. Magn. Mat. **38**, 273 (1983).

XIII.3 C. Liu, E. R. Moog and S. D. Bader, Phys. Rev. Lett. **60** , 2422 (1988).

XIII.4 G. Bergmann, Phys. Rev. Lett. **41**, 264 (1978).

XIII.5 See, for example, B. Heinrich, K. B. Urquhart, A. S. Arrott, J. F. Cochran, K. Myrtle and S. T. Purcell, Phys. Rev. Lett. **59**, 1756 (1987).

XIII.6 H. C. Siegmann, F. Meier, M. Erbudak and M. Landolt, Adv. Electronics and Electron Phys. **62**, 1 (1984).

XIII.7 W. Durr, M. Taborelli, O. Paul, R. Germar, W. Gudat, D. Pescia and M. Landolt, Phys. Rev. Lett. **62**, 206 (1989).

XIII.8 M. Lui, J. Drucker, A. R. King, J. P. Kotthaus, P. K. Hansma and V. Jaccarino, Phys. Rev. B**33**, 7720 (1986).

XIII.9 W. H. Meiklejohn and C. P. Bean, Phys. Rev. B **102**, 1413 (1959).

XIII.10 W. H. Meiklejohn, J. Appl. Phys. **33**, 1328 (1962).

XIII.11 L. Néel, Ann. Phys. **2**, 61 (1967).

XIII.12 C. Schlenker, Phys. Status Solidi **28**, 507 (1968).

XIII.13 E. Fulcomer and S. H. Charap, J.Appl. Phys. **43**, 4184 and 4190 (1972).

XIII.14 A. Yelon, in *Physics of Thin Films*, edited by M. Francombe and R. Hoffmann. Academic Press, New York, 1971. Vol. 6, p 205.

XIII.15 R. D. Hempstead, S. Krongelb and D. A. Thompson, IEEE Trans. Mag. **14**, 521 (1978).

XIII.16 C. Tsang and K. Lee, J. Appl. Phys. **53**, 2605 (1982).

XIII.17 O. Kubaschewski, *Iron–Binary Phase Diagrams*, Springer–Verlag, Berlin, 1982.

XIII.18 See, for example, C.W. Chen, *Magnetism and Metallurgy of Soft Magnetic Materials*, Dover, New York, 1977.

XIII.19 V. S. Speriosu, S. S. P. Parkin and C. H. Wilts, IEEE. Trans. Mag. **MAG–23**, 2999 (1987).

XIII.20 W. Stoecklein, S. S. P. Parkin and J. C. Scott, Phys. Rev. B **38**, 6847 (1988).

XIII.21 S. S. P. Parkin, D. P. Brunco and V. S. Speriosu, Bull. Amer. Phys. Soc. **32**, 802 (1987) and D. P. Brunco, M.I.T. Course III–B Co–Op Report, September 1986 (unpublished).

XIII.22 D. Mauri, E. Kay, D. Scholl and J. K. Howard, J. Appl. Phys. **62**, 2929 (1987); and *ibid.* **62**, 4190 (1987).

XIII.23 U. Gradmann, R. Bergholz and E. Bergter, Thin Solid Films **126**, 107 (1985).

XIII.24 M. Stampanoni, Dissertation, ETH 8937, Zurich (1989).

XIII.25 H. B. Callen and E. Callen, J. Phys. Chem. Solids **27**, 1271 (1966).

XIV. Lattice Mismatched Magnetic Superlattices

I.K. Schuller

Physics Department, B019, University of California, San Diego,
La Jolla, CA 92093, USA

This is a brief review of our work on lattice mismatched magnetic superlattices. A number of physical phenomena at different length scales have been studied using lattice mismatched magnetic multilayered structures. These include thin film, interfacial, proximity, coupling, and superlattice effects in a variety of magnetic systems.

XIV.1. Introduction

Multilayered structures have been used for many years to study a large number of physical phenomena [XIV.1–XIV.3]. In general, phenomena to be studied can be categorized according to the length scale that determines the physics. "Long" length scale phenomena (above 100 Å) include low–temperature superconductivity, electric and magnetic dipolar coupling, whereas "short" length scale studies (below 100 Å) have concentrated on crystallization, magnetic coupling (RKKY, spiral magnetism), transport and epitaxial studies. Preparation and structural characterization have received considerable attention since in many cases the physical phenomena being studied require control of layer thicknesses at the atomic level.

The effects observed can be categorized in increasing order of complexity (according to the number of layers) as: thin film, dimensional, interfacial, proximity, coupling, and superlattice effects. **Thin film** effects can, in principle, be observed in single thin films; multilayers are collections of individual films. **Dimensional** effects occur because the thickness of the layers are smaller than a characteristic length in the problem. **Interfacial** effects are observable principally because of the presence of an interface between two dissimilar materials. **Proximity** effects occur due to the effect that one material has on another across an interface. **Coupling** effects can be studied in general if two like–layers are coupled across an unlike layer. The effects mentioned above, in principle, only require the study of a single, double, or triple layer. However, due to a number of technical reasons (contamination, signal size, etc.) it is more convenient to study these phenomena in multilayered structures, which does not require *in–situ* studies. Superlattice effects, on the other hand, intrinsically rely on the periodic nature of a superlattice and therefore cannot be observed in a few layers. A variety of superlattice effects have been predicted, including the observation of phonon folding, changes in the optical properties due to electric dipolar coupling, changes in the electronic density of states in low–temperature superconducting superlattices, nonlinear I–V characteristics in metal/metal superlattices and the development of collective magnon bands in magnetic/normal superlattices. To the best of my knowledge, only phonon folding in GaAs/AlAs superlattices and collective magnon bands in Mo/Ni superlattices have been observed. I will describe here some of our work in this field, selecting representative examples from some of the categories described above.

XIV.2. Preparation and Characterization

Two principal methods of preparation have been used to manufacture multilayers: sputtering and thermal evaporation. Thermal evaporation in ultrahigh vacuum on heated substrates has been commonly designated as molecular beam epitaxy (MBE). The main advantage of sputtering is that it allows tuning the energy of particles at the substrate by changing sputtering pressure or target–substrate distance [XIV.4]. In addition, large volumes of sample can be prepared which can be used for low–sensitivity measurements or in applications requiring large areas, for instance X–ray optics. The main disadvantage of sputtering is the contamination produced by the sputtering gas and the fact that *in–situ* monitoring of growth is not possible. Thermal evaporation, especially in a UHV environment, on the other hand, permits *in–situ* monitoring of the growth and minimizes contamination. However, in general only small volume samples are produced and the maximum areas are limited. Both preparation methods are complementary. It is therefore interesting to compare structure and physical properties of samples produced by both methods.

XIV.3. Lattice Matching and Phase Diagram

It is commonly accepted that in order to improve the crystallographic quality of a superlattice, lattice matching is a necessary and sufficient condition. However, it is important to note that in many cases, systems that are lattice matched and have the same crystal structure, also form solid solutions in their thermodynamic phase diagram. In order to obtain atomically sharp interfaces, growth kinetics has to play an extremely important role to limit diffusion at the atomic level. In most case, however, the samples are grown at elevated enough temperatures, which makes interdiffusion likely. It is possible, therefore, that the thermodynamic phase diagram of the constituents plays a more important role than hitherto realized [XIV.5].

In order to avoid interdiffusion, our approach has been to concentrate on systems that do not exhibit any solid solutions in their thermodynamic phase diagram. As a consequence, the systems described here are lattice mismatched.

XIV.4. Structural Characterization

Structural characterization at the atomic level is an issue of major concern which has not been highlighted sufficiently. Theoretically, physical phenomena which occur at short length scales are strongly affected by a variety of structural defects which occur at the atomic level. These include interfacial roughness, interdiffusion, dislocations, stacking faults, etc. [XIV.6,XIV.7] To the best of my knowledge it is very difficult to ascertain, using one single technique, integrity of an interface at the atomic level. Many of the surface type probes such as Auger Electron Spectroscopy (AES), X–ray Photoelectron Spectroscopy (XPS), Low Energy Electron Diffraction (LEED), etc. are sensitive to electrons arriving from a finite escape depth from the samples. Scattering techniques, such as electron, X–ray or neutron diffraction, require extensive modelling which only recently is being addressed in detail [XIV.8]. Other structural probes, such as High Energy Electron Diffraction (HEED), although only sensitive to the first monolayer, have only been used to extract qualitative information and are not sensitive to chemical intermixing.

Clearly, only a comprehensive approach using a variety of techniques, together with quantitative modelling, will allow solving structural issues at the atomic level. I believe that in the near future more and more effort will be dedicated to issues related to structure at the atomic level.

It should be pointed out that a number of physical phenomena can be studied at large–length scale where structural defects at the atomic scale do not play a major role [XIV.9].

XIV.5. Magnetic Properties

The magnetic properties of lattice mismatched metallic superlattices exhibit a variety of interesting phenomena at short and long scales [XIV.9,XIV.10]. In general, the phenomena at short scales may be caused by unusual interfacial physics. However, the possibility of interfacial structure affecting the results has not been uniquely ruled out. These effects include the decrease in magnetization [XIV.11] and changes in first–order anisotropy in short wavelength Mo/Ni superlattices [XIV.12]. An interesting phenomenon at long scales (due to dipolar coupling) is related to the development of magnon bands in Mo/Ni superlattices [XIV.13].

a) Thin–film and interfacial effects

The saturation magnetization of Mo/Ni superlattices show systematic trends which are characteristic of thin–film [XIV.11] effects. Fig. XIV.1 shows the low–temperature saturation magnetization of Mo/Ni superlattices as a function of Ni thickness for three series of samples. Generally all the data fall close to a universal curve with the magnetization decreasing with Ni thickness. It is interesting

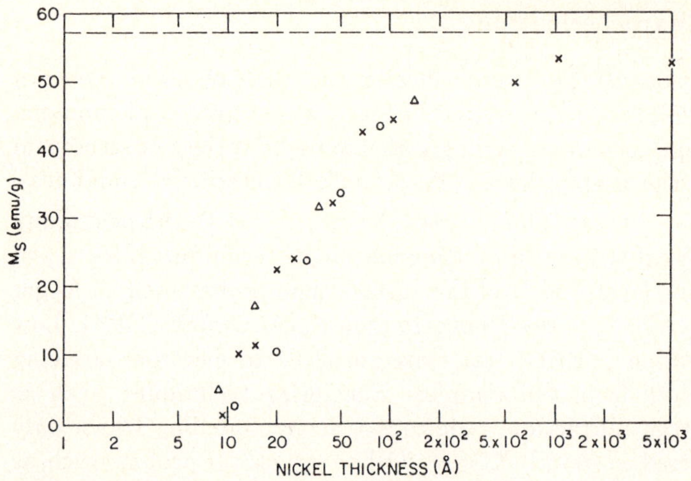

Figure XIV.1. Saturation magnetization as a function of nickel thickness for $\square$, $D_{Ni} = 3D_{Mo}$; $\times$, $D_{Ni} = D_{Mo}$; $\circ$, $D_{Mo} = 3D_{Ni}$.

to note that the curves with thinner Mo have a systematically higher saturation magnetization [XIV.9]. This type of behavior is typical of thin Ni films and is thought to arise from the fact that Ni atoms on the interfaces have fewer Ni nearest neighbors than the ones in the bulk. As a consequence, the (presumably) nonmagnetic interfacial Ni atoms become progressively more important in determining the *average* magnetization with decreasing layer thickness. The enhancement of the magnetization for thinner Mo layers could be due to a variety of effects; percolation ("shorts across the molibdenum"), or some form of magnetic coupling (RKKY, exchange, etc.) across the nonmagnetic molybdenum. This enhancement points out the need for clear–cut structural studies capable of establishing the perfection of the intervening nonmagnetic Mo layers. The X–ray diffraction data in this system indicates the presence of 1–2 atomic–planes roughness at an interface. Therefore, it is thought that the coupling effects arise because of percolation and not because of some more exotic coupling across a normal metal.

One interesting magnetic phenomenon at short scale is observed in the interfacial anisotropy [XIV.12]. Fig. XIV.2 shows a graph of the first–order anisotropy $(H_a^{(1)})$ as a function of Ni thickness measured by two different methods: DC magnetization and ferromagnetic resonance (FMR). To extract the first–order anisotropy from the DC magnetization, a fit to first– and second–order anisotropy was made, as shown in the inset to Fig. XIV.2. The first order anisotropy is in good agreement at large Ni thicknesses. However, for small thicknesses, there is a serious disagreement between the two types of measurements. This disagreement has been shown to scale inversely with the Ni thickness and therefore it is thought to arise from a surface contribution. At

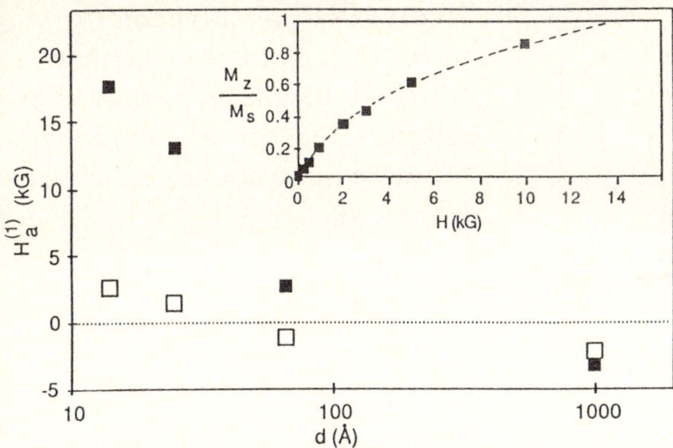

Figure XIV.2. Low–temperature, first–order anisotropy vs Ni–layer thickness for FMR (solid squares) and DC magnetization (open squares). The dotted line marks the transition from planar to perpendicular H_a. Inset: DC magnetization of Mo(25 Å)/Ni(25 Å) sample, compared to higher–order anisotropy model (dashed line).

this stage we should stress that irrespective of the details of the structure, the two types of measurements should give the same results. At present, the origin of this discrepancy has not been clarified. It may arise from an incomplete understanding of the measurements or maybe due to an enhanced anisotropy sensed and/or perhaps caused by the FMR experiment.

b) Superlattice effect

Superlattice effects rely on the existence of an additional periodicity imposed on the lattice by the superstructure. These types of effects cannot be observed even in principle in a few layers and require the additional superlattice periodicity.

The type of superlattice effects described here rely on long–range dipolar coupling. For this type of effects, structural perfection at atomic length scales is not crucial. The results of theoretical calculations [XIV.14,XIV.15] are schematically shown in Fig. XIV.3, where the frequency of magnon modes (ν) is plotted as a function of normal metal thickness (t_{normal}). An isolated thin film exhibits one "surface" mode whereas in the bulk material two modes are present, a surface and a bulk modes. In the transition region, a band of magnons is present. Theoretical calculations show that the coupling of light to the magnons has a higher density of states at the bottom of the band. The arrows show three series of Mo/Ni superlattice samples for which the magnon spectrum was measured, using Brillouin scattering techniques [XIV.13].

Fig. XIV.4 shows the frequency of the magnons for representative samples from the three series, together with fits to the theoretical models. The saturation magnetization extracted from these measurements are within 15% of independent measurements using DC magnetization measurements (see Fig. XIV.1).

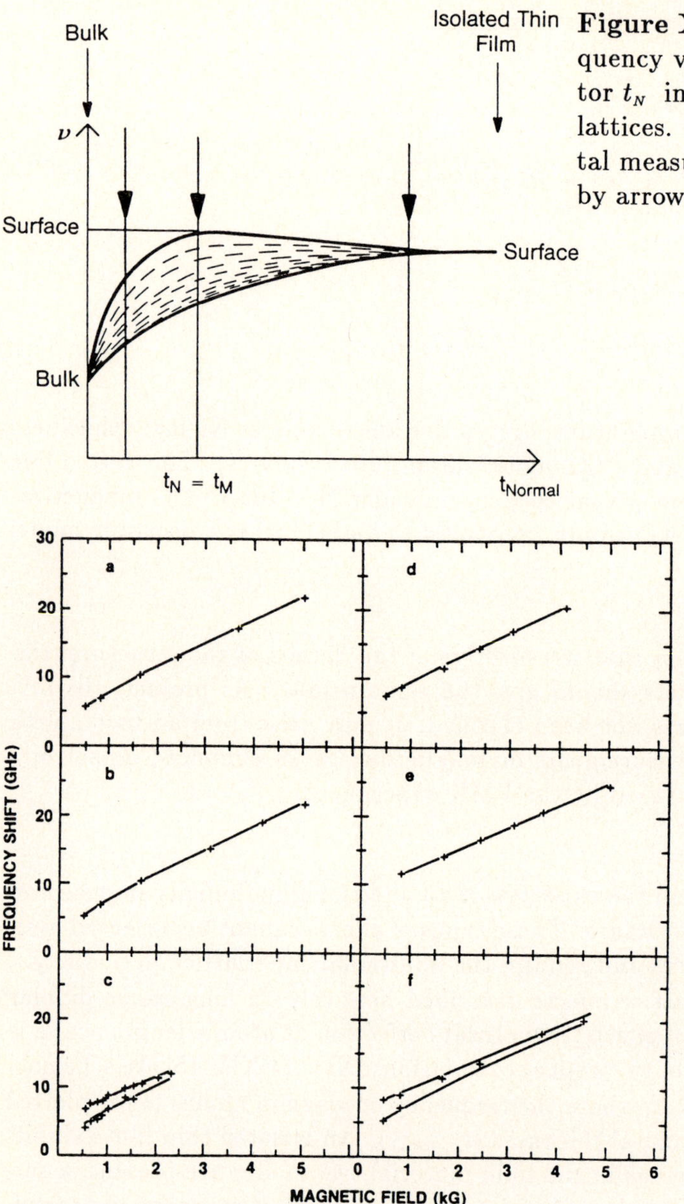

Figure XIV.3. Magnon frequency vs. normal metal separator t_N in magnetic/normal superlattices. The series of experimental measurements are indicated by arrows.

Figure XIV.4. Magnon frequency (crosses) for representative Mo/Ni superlattices *vs.* magnetic field together with theoretical fits (solid lines).

Measurements as a function of light scattering wavevector are also in quantitative agreement with the theoretical calculations. The excellent quantitative agreement between theoretical calculations and experimental measurement shows that the understanding of these phenomena is on firm footing.

XIV.6. Summary

Magnetic superlattices serve as ideal model systems to study physical phenomena in reduced dimensionality and to engineer novel properties. We have emphasized that phenomena which occur at short scales, below 100 Å require the preparation and characterization of samples with sharp, well segregated interfaces at atomic length scales. It is expected that future efforts will be dedicated to a thorough understanding of structure at the atomic scale. Several interesting applications were described which do not require atomic–scale structural perfection.

I thank my collaborators over several years and the Office of Naval Research for support in the initial stages of this work. Work supported by DOE grant #DE–FG03–87ER45332.

References

XIV.1 *"Synthetic Modulated Structures"*, L. L. Chang and B. C. Giessen eds., Academic Press, Inc., Orlando (1985).

XIV.2 *"Interfaces, Superlattices and Thin Films"*, J. D. Dow and I. K. Schuller eds., Materials Research Society Publishers, Vol. 77, Pittsburgh, PA (1987).

XIV.3 *"Physics, Fabrication and Applications of Multilayered Structures"*, P. Dhez and C. Weisbuch eds., Plenum Publishing Co. (1988).

XIV.4 K. E. Meyer, I. K. Schuller, and C. M. Falco, J. Appl. Phys. **52**, 5803 (1981).

XIV.5 I. K. Schuller, Superlattices and Microstructures **4**, 521 (1988).

XIV.6 See, for instance, D. B. McWhan in Ref. 1, pg. 43.

XIV.7 See, for instance, D. B. McWhan in Ref. 2.

XIV.8 See, for instance, J. P. Locquet, D. Neerinck, W. Sevenhans, Y. Bruynseraede, H. Homma, and I. K. Schuller in *"Multilayers; Synthesis, Properties and Non–Electronic Applications"*, T. W. Barbee Jr., F. Spaepen and L. Greer eds., Materials Research Society, vol. 103 (1988), pg. 217.

XIV.9 See, for instance, I. K. Schuller and H. Homma, Materials Research Society Bulletin **XII**, 18 (1987).

XIV.10 For a review, see I. K. Schuller in AIP Conf. Proc. **138**, 93 (1986).

XIV.11 M. R. Khan, P. Roach, and I. K. Schuller, Thin Solid Films **122**, 183 (1985).

XIV.12 M. J. Pechan and I. K. Schuller, Phys. Rev. Lett. **59**, 594 (1987).

XIV.13 A. Kueny, M. Khan, I. K. Schuller, and M. Grimsditch, Phys. Rev. B **29**, 4915 (1984).

XIV.14 P. Grünberg and K. Mika, Phys. Rev. B **27**, 2955 (1983).

XIV.15 R. E. Camley, T. S. Rahman, and D. Mills, Phys. Rev. B **27**, 261 (1983).

XV. Anisotropic Magnetic Response of Rare Earths in Superlattices

C.P. Flynn[1], *F. Tsui*[1], *M.B. Salamon*[1], *R.W. Erwin*[2], *and J.J. Rhyne*[2]

[1]Department of Physics and Materials Research Laboratory,
 University of Illinois at Urbana-Champaign, 104 S. Goodwin,
 Urbana, IL 61801, USA
[2]National Institute of Standards and Technology,
 Gaithersburg, MD 20899, USA

We have observed that Dy layers, in Dy/Y superlattices grown along the **c** axis, couple together to produce long–range order even through Y space layers over 120 Å thick. In superlattices grown along the **b** axis, however, no coupling is observed in Y space layers as little as 26 Å thick. The observed range and anisotropy of the interaction is discussed in terms of the fundamental response Y to local magnetic perturbations.

XV.1. Introduction

A number of publications over the past four years have reported properties of rare earth superlattices grown with high structural quality by molecular beam epitaxy [XV.1,XV.2]. The greatest progress has been made with structures in which layers of a magnetic rare earth alternate with layers of nonmagnetic Y. Both ferromagnetic rare earths like Gd [XV.1], and those exhibiting antiferromagnetic order in the form of linear spin waves [XV.2] (e.g., Er) and helical spin waves [XV.2] (e.g., Dy), have been employed as the magnetic component [XV.3]. Superlattices containing instead two alternating magnetic components have also been prepared, but their properties appear rather too complicated to be interpretable at this time [XV.1]. The initial choice of Y as a nonmagnetic spacer was made because Y is a fortunate and interesting choice for reasons of magnetic behavior, in addition to its structural merits. The present paper describes recent progress in probing the magnetic response of Y to local magnetic perturbations, and in understanding the way this response determines the properties of superlattices in which magnetic components alternate with Y space layers.

Two principal factors arise in the magnetic structure of rare earth superlattices. The first of these is the modified behavior of the magnetic component, and the second is the coupling among magnetic layers through the nonmagnetic spacers. Rare earth magnetism is a consequence of quite localized electronic interactions among spins (the mean free path is in any event short). Therefore the magnetic structure is not greatly modified by partitioning into layers, even down to 10 atoms thick, other than for demagnetization effects. Rather, the principal effect of superlattice geometry on the behavior of magnetic layers is due to "clamping" caused by epitaxy. In effect, the dimensions of the magnetic

Springer Proceedings in Physics, Vol. 50   **Magnetic Properties of Low-Dimensional Systems II**
Editors: L.M. Falicov · F. Mejía-Lira · J.L. Morán-López © Springer-Verlag Berlin, Heidelberg 1990

layer in the growth plane are locked to those of the substrate, and the magnetic
behavior is modified accordingly. For example, the rare earths Dy and Er in
bulk form exhibit large magnetostrictive strains upon turning ferromagnetic;
this clamping turns out to completely suppress the ferromagnetic transitions in
thin films and superlattices containing Dy [XV.4] and Er [XV.5]. It also causes
the wavelengths of the antiferromagnetic phases to differ from those in the bulk.
These striking effects are the most significant changes of magnetic order within
the magnetic layers that arise in superlattices and buried thin layers. The remaining order in superlattices, perpendicular to magnetic layers, originates from
their interactions through the Y spacers. It provides the main subject matter
in what follows.

Experiments on superlattices quickly revealed that magnetic layers can couple through "nonmagnetic" Y. They do so at remarkably large separations of
100 Å or more. The mechanism can be pinned down with some certainty as the
RKKY oscillatory disturbance of the conduction states by the magnetic ions
[XV.6,XV.1]. In Gd/Y superlattices, for example, the coupling energy varies
with spacer width in an apparently oscillatory manner and with roughly the
predicted wavelength. Even more striking is the behavior of Dy/Y systems
[XV.4] where *chirality* of the Dy magnetism is observed to order from one layer
to next. The helical structure apparently gives rise to two orthogonal responses
that necessarily are *out of phase*, as sketched in Fig. XV.1. It is apparent that
this preserves chirality in the superposed oscillatory responses, and the order is
thus carried from one layer to the next. These observations leave little doubt
that it is the RKKY interaction rather than structural flaws, for example, that
cause the observed order among layers. Other evidence will emerge in Section
XV.3.

It is useful to consider a single magnetic ion in Y. In response to this magnetic perturbation the conduction electrons polarize to produce "Friedel wiggles"
and these in turn cause the RKKY interaction between magnetic ions. The spatial form of this response is of some interest. One may regard the superlattice

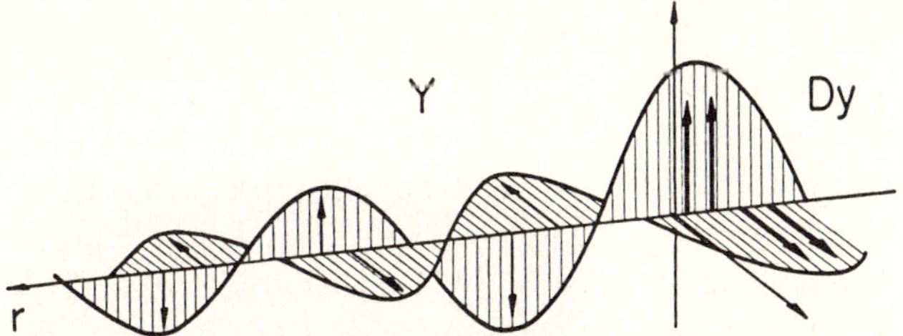

Figure XV.1. Spiral magnetism (heavy arrows, right) stimulates two orthogonal RKKY responses that are out of phase. The chirality is seen to be preserved
in the RKKY response, which transmits the structure to the next magnetic
layer.

experiments described above as establishing that the response remains substantial over a range $\geq 100\text{Å}$ that is, perhaps surprisingly long. This range pertains to the hexagonal **c** axis, as the superlattices under discussion were grown on their basal planes perpendicular to the **c** axis. One naturally enquires what the range may be along other directions. What is the three–dimensional character of the response? Are these measured properties in agreement with theoretical predictions? The necessary measurements along the **a** and **b** axes require entirely new growth procedures. These are described in Section XV.2, and Section XV.3 presents the results of structural and magnetic measurements. The degree to which the results can be reconciled with the theory is the subject matter of the Discussion in Section XV.4.

XV.2. Crystal Growth

Two discoveries opened the path to growth of **c** axis rare earths. First, *bcc* Nb(110) grows on sapphire $(11\bar{2}0)$ at about 900 C as an excellent single crystal [XV.7]. Second, Y and hexagonal rare earths grow in their (0001) planes on Nb(110), at temperatures of 400 C or a little higher [XV.8]. These rare earths can be grown on each other and on Y with few if any problems, and a Y cap serves to prevent subsequent oxidation in air. In the most favorable cases the

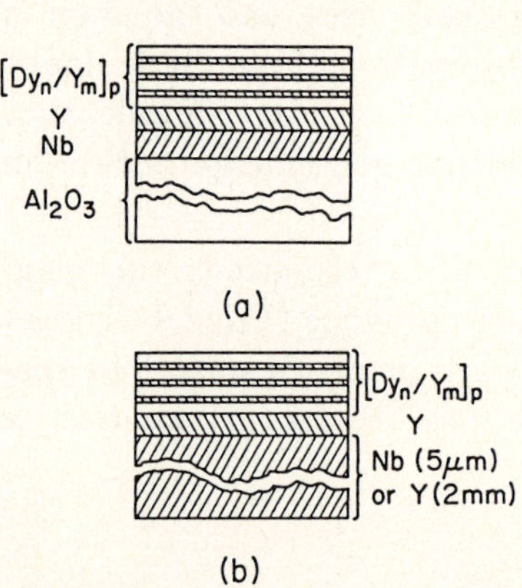

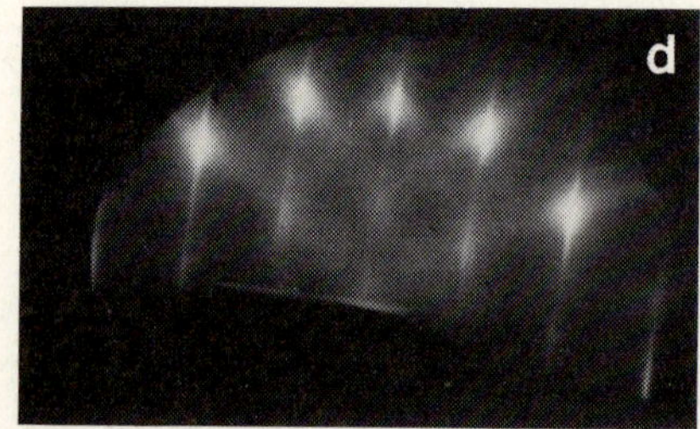

Figure XV.2. Sketch of Dy/Y superlattice structures for **c** axis samples grown on Al$_2$O$_3$ (a) and **b** axis samples on single crystal Y (b). Bragg reflections in (c) from **b** axis sample with 87 Å wavelength and 40 periods show clear Cu $K_{\alpha 1} - K_{\alpha 2}$ splittings, arising from high structural coherence. (d) shows circular pattern of RHEED spot from an excellent flat Dy surface grown by MBE.

rare earth surface can be extremely flat so that the RHEED streaks contract to leave mainly the circular pattern of 2D diffraction spots where the surface diffraction rods intersect the Ewald sphere. This assures the high quality of superlattice interfaces in carefully grown samples. The overall structure of **c** axis samples is sketched in Fig. XV.2a.

It is generally more difficult to grow hexagonal materials along axes other than **c**, and the rare earths are no exception. Our procedures [XV.9] start from Y single crystals grown at the Ames Laboratory with either the $(11\bar{2}0)$ (i.e., **a**) or $(1\bar{1}00)$ direction (i.e., **b**) perpendicular to the polished face. After a cleaning by sputter–anneal cycling in the uhv chamber of the MBE machine single crystal Y can be grown along either the **a** or **b** axis at choice. The surfaces tend to be poorly reconstructed during growth but X–ray analysis reveals that the final structure of the buried crystal is excellent. By precisely these same methods is possible to grow **a** and **b** axis superlattices. The resulting overall structure is sketched in Fig. XV.2b. An X–ray diffraction Bragg scan of **b** axis sample is shown in Fig. XV.2c. A RHEED scan of an excellent basal plane Dy surface is reproduced in Fig. XV.2d.

XV.3. Magnetic Structure

The behavior of rare earth superlattices is interesting in a number of respects, including the long– and short–range magnetic order, their temperature dependences, the existence of field–induced ferromagnetism and the supression of phase transitions by lattice clamping. For brevity we shall focus here mainly on long–range order in the antiferromagnetic phase of Dy layers of Dy/Y super-lattices. The order has been investigated by neutron scattering methods on a triple axis spectrometer at the National Institute of Standards and Technology Reactor. Fairly extensive investigations have been completed on a number of **a**, **b** and **c** axis samples. It appears that **a** and **b** axis samples behave quite similarly, so a comparison of **c** and **b** axis is sufficient here. The main points are summarized in what follows.

Both **b** and **c** axis Dy layers undergo transitions [XV.4,XV.11] to the antiferromagnetic (spiral) phase close to the bulk Néel temperature $T_N = 178$ K. In both geometries the bulk transition to the ferromagnetic phase at 78 K is completely suppressed in thin films, presumably owing to the effect of lattice clamping on the magnetoelastic energy. This transition is recovered at low temperatures in films $\geq 10^2$ Å thick (Er films exhibit no ferromagnetism even up to 0.5μ thickness) [XV.1,XV.5]. In Dy **c** axis samples the magnetic layers couple to form an ordered structure that is coherent over several superlattice periods, even close to T_N, and for space layers over 120 Å wide [XV.4,XV.10]. In **b** axis Dy samples, however, no magnetic order can be detected among the Dy layers down to 4 K even when the Y spacers are only 26 Å wide [XV.11]. This major difference and its explanation are the main results described in the present work.

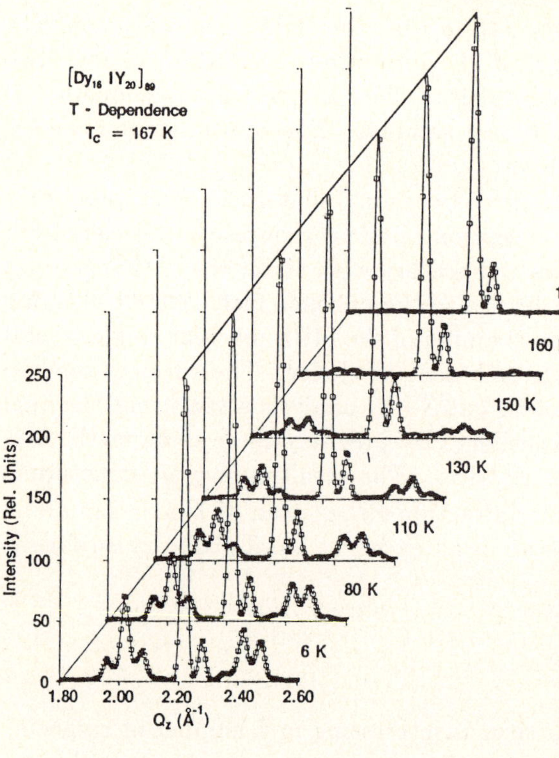

Figure XV.3. Neutron diffraction scans along [000l] in **c** axis [Dy16/Y$_{20}$]$_{89}$ for several temperatures. Note the temperature independence of the [0002] peak at $Q_z = 2.215\text{Å}^{-1}$. The small peak to the right of [0002] is a bilayer harmonic. The fundamental and two bilayer harmonics are shown for both Q (≈ 2.42) magnetic satellites and are observed to be temperature dependent.

Fig. XV.3 shows a neutron diffraction scan [XV.10] along (000l) for **c** axis [Dy$_{16}$/Y$_{20}$]$_{89}$ at several temperatures. Here the subscripts on the chemical symbols indicate the numbers of basal plane layers of each component per period and the final subscript gives the number of complete periods. Several features of the spectra require comment. One instrumental point is that the linewidths are limited by the instrumental resolution. In reality, the structural coherence exceeds 500 Å, and thus includes many superlattice periods.

The overall form of the Bragg scans in Fig. XV.3 may be analyzed as follows. There is a large central peak that corresponds to Bragg scattering from the periodic arrangements of the nuclei. Separated from this peak by $\sim 0.2\,\text{Å}^{-1}$ on either side are magnetic side bands. As in bulk Dy, [XV.3] the atomic moments are aligned in each basal plane, but their alignment axis changes systematically by about 60° each two planes, and with a given helicity. In this way a new (incommesurate) periodicity is introduced, to which the neutron diffraction responds. This interpretation is borne out by the observed intensity decrease of the satellites (accurately following a Brillouin function) as the temperature approaches the Néel point $T_N \simeq 175$ K. Both the magnetic satellites and the central diffraction are broken into series of peaks spaced by about 0.062 Å^{-1}. That these are structural sidebands due to the superlattice periodicity is established by the observed period L=102 Å, for which $2\pi/\text{L}=0.0616\ \text{Å}^{-1}$. Careful scrutiny reveals that the magnetic satellites are separated from the central peak by a splitting that is temperature dependent. The wavelength of the magnetic spiral and its temperature dependence both differ from those in the bulk. They depend in addition on superlattice characteristics including the magnetic layer thickness and superlattice wavelength. Similar shifts generally occur in thin buried layers. The main cause is the lattice clamping by the epitaxial constraint, as mentioned above.

For the present purposes the most important information revealed by Fig. XV.3 is that the helicity propagates from one Dy layer to the next coherently over many superlattice periods. The width of the peaks measures the magnetic coherence length. Corrections are of course needed for instrumental broadening, which leave the lines still narrower. Even uncorrected, however, the linewidths are less than the superlattice splitting, which means that the coherence length exceeds the superlattice wavelength. Accurate fits reveal coherence up to five or six superlattice periods at low temperatures. The coherence length is much reduced at temperatures close to T_N. It may also be reduced by fields applied in the basal plane that tend to produce ferromagnetic alignment.

From the long coherence length it is possible to deduce in a fairly convincing way that successive Dy layers interact through RKKY disturbances of the intervening Y. If successive Dy layers had random chiralities the phase of the magnetic spiral along **c** would perform a random walk and the coherence indicated by its Fourier transform would just be the Dy layer thickness. The fact that the coherence length is, to the contrary, an order of magnitude longer shows unambiguously that neighboring layers exhibit identical chiralities. This in turn establishes that the interaction through the Y is large enough to produce the ordering. Ordering of this type is observed for Y spacers over 120 Å wide. Fig. XV.1 illustrates the mechanism by which the chiral information is transmitted through two orthogonal RKKY responses. Actual oscillations of coupling strength with layer separation have been reported for Gd/Y superlattices, in which the magnetism is confined to one orientation, so that the RKKY response oscillates [XV.1] with linear polarization.

We now turn to superlattices grown along the **b** axis $(1\bar{1}00)$, for which fewer data are, as yet, available. Neutron scattering measurements [XV.11] show that the **c** axis spiral again forms, as in the bulk, even for Dy layers only 7 planes thick (20 Å). As in the **c** axis case the Néel temperature is lowered, as much as 10% for thin layers, the turn angle has a different temperature dependence, and the ferromagnetic transition is once more completely suppresed by lattice clamping. In contrast to the **c** axis case, however, **b** and **a** axis Dy/Y samples exhibit no sign of ordering between successive magnetic layers. Specifically, the **b** axis Dy layers fail to order along **b** even for Y spacers ~ 20Å wide, whereas order along **c** is observed for **c** axis samples with Y layers 120 Å wide. These results establish that the coupling through Y is highly anisotropic, with the **c** axis effect much stronger than that for **a** and **b** axis superlattices.

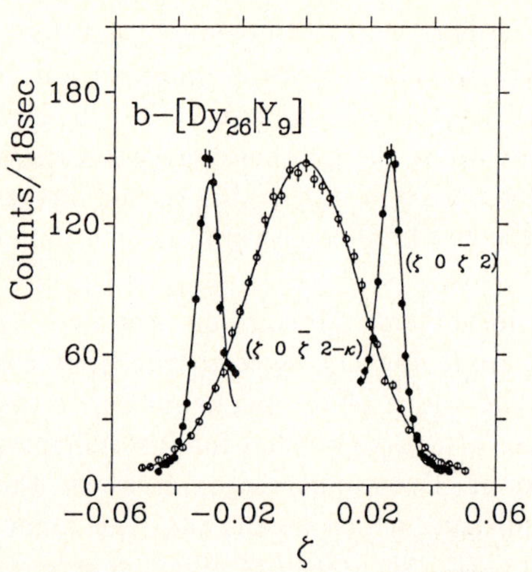

Figure XV.4. Magnetic and structural peaks of the b–axis superlattice [Dy$_{26}$/Y$_9$]$_{82}$. Open circles: scan along a* across magnetic peak at [000(2 − κ)]; closed circles: scans though [0002] along a* in a field of 2.5T to show the structural satellites of [0002].

Fig. XV.4 gives examples of data that support these conclusions [XV.11]. For **b** axis [Dy$_{26}$/Y$_9$]$_{82}$ the figure shows a scan through the magnetic satellite at [000(2−κ)] along the reciprocal lattice direction a* (i.e. perpendicular to the growth plane) with ς the scan variable. For this sample at 6 K the magnetic splitting q$_0$ $=\kappa$c*$=0.227$ Å^{-1}. The scan along a* should reveal the degree to which coherent magnetic structure is present in this direction. In fact, however, the peak width corresponds to a coherence length of only 80 Å. This is indistinguishable from the 73 Å thickness of individual Dy layers. Thus there is negligible coupling of successive Dy layers along **b** when separated by just 26 Å of Y. Note that the line width in Fig. XV.4 is, in fact, necessarily associated with magnetic incoherence rather than structural disorder. This is established by the sharp structural satellites that become visible (Fig. XV.4) when a 2.5 Tesla magnetic field is applied to saturate the material along the easy (**a** axis) direction of magnetization. Their widths are ~ 0.017Å^{-1} as compared to the resolution of 0.010 Å^{-1}, whereas the scan for B=0 in Fig. XV.4 has a width of 0.087 Å^{-1}. Interestingly enough the coherence along **a** in the growth plane is long, ≥ 500 Å, whereas that along **c** is considerably shorter, perhaps owing to a position–dependence of the turn angle. Space limitations preclude further discussion of these phenomena here.

XV.4. Discussion

A framework for the discussion of magnetic response in rare earth metals has been developed over the past two decades [XV.6]. Our new abilities to structure materials on an atomic scale opens opportunities for exploring the experimental consequences of the theoretical predictions, and for consequent refining of the present relatively primitive understanding of the magnetic response. In this paper we are concerned only with the coarsest aspects of the response, namely its general range and anisotropy.

In second–order perturbation theory, (i.e., linear–response theory) the conduction electrons in rare earth metals react to small magnetic perturbations with a $\mathbf{q}$–dependent linear susceptibility $\chi(\mathbf{q})$ [XV.6,XV.12,XV.13]. Two magnetic ions with spins $\mathbf{S}_1$, $\mathbf{S}_2$ at $\mathbf{r}_1$ and $\mathbf{r}_2$, are then coupled by a Hamiltonian $H = \mathbf{S}_1 \cdot \mathbf{S}_2 \, J(\mathbf{R})$, in which $\mathbf{R} = \mathbf{r}_2 - \mathbf{r}_1$, and

$$J(\mathbf{R}) = \sum_q j(\mathbf{q}) e^{-i\mathbf{q}\cdot\mathbf{R}} , \qquad\qquad (XV.1)$$

with $j(\mathbf{q}) = |j_{sf}(\mathbf{q})|^2 \, \chi(\mathbf{q})/2$ and $j_{sf}(\mathbf{q})$ the exchange matrix element connecting band states to the 4f core [XV.6,XV.12]. For an isotropic electron gas $J(\mathbf{R})$ takes the well known RKKY form $\sim \cos(2k_F R)/(k_F R)^3$, but in the present application the particular spatial form of $J(\mathbf{R})$ plays a central role. The interaction must be summed over all pairs to obtain the energy. Detailed electronic processes such as thermal or spin scattering, or the sharing of a common Fermi surface among the layers of a superlattice, enter only through their effect on $J(\mathbf{R})$. This approach remains valid for small changes about the mean, even when the summed interactions exceed the linear regime. When perturbations are clustered, for example, so that the material is no longer homogeneous, linear–response theory still holds, but $J(\mathbf{r}_1,\mathbf{r}_2)$ depends explicity on $\mathbf{r}_1$, $\mathbf{r}_2$ rather than $\mathbf{R} = \mathbf{r}_2 - \mathbf{r}_1$ alone. In the superlattices of interest here, however, the actual differences of $\chi(\mathbf{q})$ in Eqn. (XV.1) between Y and Dy are known to be quite small. The effect of inhomogeneity may therefore be neglected in the following approximate description.

For Dy and Y in the superlattices discussed here, it is well documented that $j(\mathbf{q})$ has peaks along $\mathbf{c}^*$ in reciprocal space that arise from almost two–dimensional sheets of nesting hole Fermi surface. Along $\mathbf{a}^*$ and $\mathbf{b}^*$, $j(\mathbf{q})$ falls off in a way that mainly reflects the core size, ans may be represented by a gaussian of width 0.63 Å^{-1} (a faster fall may occur when $\mathbf{q}$ exceeds the span of the hole Fermi surface, but this is not significant here). The peaks along $\mathbf{c}^*$ are difficult to model accurately. They differ somewhat between Y and Dy, and even more among different theories for Y. Liu et $al.$ [XV.13] find a peak $\sim 0.7j(0)$ high with a markedly square top and structure ~ 0.03Å^{-1} wide. The general features of $j(\mathbf{q})$ are sketched in Fig. XV.5a. According to Eq. (XV.2) $J(\mathbf{R})$ is the Fourier transform of $j(\mathbf{q})$, and its general envelope is sketched in Fig. XV.5b, consistently with Fig. XV.5a. Also shown there is the actual $J(\mathbf{R})$ along $\mathbf{c}$ calculated from Liu's results. Earlier calculations by Kasuya [XV.12] give $j(\mathbf{q})$ whose transform $J(\mathbf{R})$ has a similar range and wavelength to Liu's, but $J(\mathbf{R})$ falls off without beats. The main features that emerge unambiguously from these calculations is that $J(\mathbf{R})$ extends with significant amplitude for some 130 Å along $\mathbf{c}$, but falls off to the same strength in only about 12 Å along $\mathbf{a}$ and $\mathbf{b}$.

Several aspects of the magnetic behavior are clarified by this picture. For our present purpose the most important one is that the anisotropy of $J(\mathbf{R})$

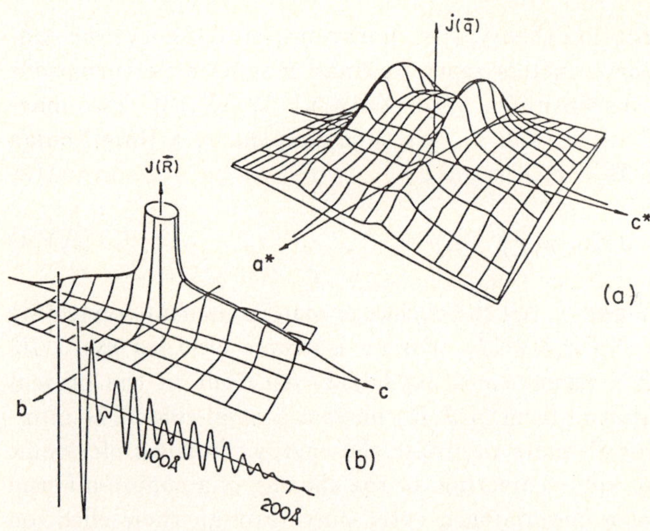

Figure XV.5. (a) Schematic representation of $j(\mathbf{q})$. The width along $\mathbf{a}^*$ is determined mainly by $j_{sf}(\mathbf{q})$ while that along $\mathbf{c}^*$ fits sharp features in $\chi(\mathbf{q})$. (b) Sketch of the envelope function of the Fourier transform of $j(\mathbf{q})$ in (a), showing the anisotropic spatial extent. Also shown is the actual oscillatory Fourier transform $J(\mathbf{R})$ of Liu's $j(\mathbf{q})$ along the $\mathbf{c}$ axis.

explains in a natural way why magnetic layers in superlattices couple at separations over 100 Å along $\mathbf{c}$ but fail to couple at 26 Å along $\mathbf{a}$ and $\mathbf{b}$. While the theory is not fully quantitative, both the ranges and the anisotropy are predicted reasonably well. Furthermore, since the susceptibility peaks and matrix elements are similar for Dy and Y, it is natural that these qualitative features should be preserved in the superlattices. We note that the simplified RKKY interaction $\sim \cos 2k_F R / (k_F R)^3$, for a free electron gas, would also predict small coupling in $\mathbf{b}$ axis Dy samples, because this effect, when summed over a layer, leads to an exponentially decaying response; however, the large observed coupling range along $\mathbf{c}$ shows clearly that the description used above, as refined for rare earths, is essential. Nor can the results be explained by accidental nodes of the coupling strength variation with Y spacing, as in Gd, because the Dy spiral does not produce nodes. Therefore the theory as presented appears qualitatively applicable. It may also be recognized that the approximate interaction volume $260\times24\times24\sim 10^5 \,\text{Å}^3$, or $> 10^3$ neighbors, readily accounts for the observed ordering in Y alloys containing only a few atomic percent of magnetic impurities [XV.14]. The lack of observed coupling along $\mathbf{a}$ and $\mathbf{b}$ lends confidence that the long range along $\mathbf{c}$ is not an artifact due to unknown structural flaws.

Much remains to be learned about these rare earth superlattices and the response behavior to which their properties must be attributed. The magnetic structure changes, for example, include differences of magnetic transitions and turn–angle characteristics that are understood rather poorly at present. The $\mathbf{b}$

axis samples possess an unusual magnetic coherence that is anisotropic [XV.11], most probably owing to confinement inside single layers along **b** and fluctuations of turn angle along **c**. Regardless of these remaining challenges, however, the general length scale and anisotropy of the response appear to be reasonably well characterized by our results and in qualitative accord with the available theoretical description. Both the theory and the experiments need further refinement. In this connection it appears possible that suitably tailored samples may open opportunities for a more complete experimental probing of the magnetic response of rare earth metals to local magnetic perturbations. the goal is to understand the spatial form of the local response. This will pemit fundamental links with the theory and, at the same time, make it possible to develop detailed predictions for the behavior of more complicated structures as superlattices.

The authors thank R. Du and J. Borchers for assistance. This work was supported by NSF–DMR 86–21616. The use of facilities supported by the MRL grant NSF–DMR–12860 is gratefully acknowledged.

References

XV.1 J. Kwo, *Thin Film Growth Techniques for Low Dimensional Structures*, edited by R. F. C. Farrow (Plenum Press, NY 1987) reviews work in her group, largely concerning naturals with Gd.

XV.2 For a recent review of work in the author's program see C. P. Flynn, J. Phys. Colloq. (in press).

XV.3 The structure of bulk rare earth metals is reviewed in *Magnetic Properties of Rare Earth Metals*, edited by R. J. Elliot (Plenum Press, London, 1972).

XV.4 R. W. Erwin, J. J. Rhyne, M. B. Salamon, F. Borchers, S. Sinha, R. Du, J. E. Cunningham, and C. P. Flynn, Phys. Rev. B **36**, 6808 (1987).

XV.5 J. Borchers, M. B. Salamon, R. Du, C. P. Flynn,R. W. Erwin, and J. J. Rhyne, Superlattices and Microstructures **4**, 439 (1988).

XV.6 R. Coqblin, *The Electronic Strucutre of Rare–Earth Metals and Alloys: the Magnetic Heavy Rare Earths.* (Academic Press, London, 1977); A. J. Freeman, in Ref. XV.3. A recent discussion of Gd/Y superlattices is Y. Yafet, J. Appl. Phys. **61**, 4058 (1987).

XV.7 S. M. Durbin, J. E. Cunningham. M. E. Mochel, and C. P. Flynn. J. Phys. F **11**, 223 (1981).

XV.8 J. Kwo, D. B. McWhan, M. Hong, E. M. Gyorgy, L. C. Feldman, and J. E. Cunningham in *Layered Structures, Epitaxy and Interfaces*, edited by J. H. Gibson and L. R. Dawson, Materials Research Society Symposia Proceedings, Vol. 237 (Mater. Res. Soc., Pittsburgh, PA, 1985).

XV.9 R. Du, F. Tsui, and C. P. Flynn, Phys. Rev. B **38**, 2941 (1988).

XV.10 J. J. Rhyne, R. W. Erwin, J. Borchers, S. Sinha, M. B. Salamon, R. Du, and C. P. Flynn, J. Appl. Phys. **61**, 4043 (1987).

XV.11 C. P. Flynn, F. Tsui, M. B. Salamon, J. J. Rhyne, and R. W. Erwin, to be published.

XV.12 T. Kasuya, in *Magnetism IIB*, edited by G. T. Rado and H. Suhl (Academic Press, NY, London, 1966) Ch. 3.

XV.13 S. H. Liu, R. P. Gupta, and S. K. Sinha, Phys. Rev. B **4**, 1100 (1971).

XV.14 J. A. Gotaas, J. J. Rhyne, L. E. Wenger, and J. A. Mydosh, J. Appl. Phys. **61**, 3415 (1987).

XVI. Studies of the Magnetic Properties of Superlattices in the Fe(110)/Ag(111) System

C.J. Gutierrez, S.H. Mayer, Z.Q. Qiu, H. Tang, and J.C. Walker

Department of Physics and Astronomy, Johns Hopkins University,
Baltimore, MD 21218, USA

Three series of high quality, well–characterized Fe(110)/Ag(111) superlattices with continuous Fe components 2, 5 and 8 monolayers (ML) thick and continuous Ag components of varying thicknesses were produced by molecular beam epitaxy (MBE). Transmission Mössbauer spectra of all of these films show that they are magnetically ordered at all measurement temperatures, and that their directions of magnetization are in–plane. The 5 and 8 ML Fe component superlattices measured between 4.2 K and room temperature show a $T^{3/2}$ temperature dependence of the hyperfine field. There is a strong variation in the temperature dependence with different thickness of the Ag bilayer in these superlattices. This suggests a magnetic interaction across thin silver barriers between neighboring iron bilayers. Mössbauer spectra of the 2 ML Fe bilayer superlattices taken between 4.2 K and 500 K generally consist of a composite of a sextet spectra together with a small central feature associated with the thermal relaxation of small Fe islands. The existence of these islands is further confirmed by measurements on samples in an external field. The 2 ML Fe superlattices all show a *linear* temperature dependence of the hyperfine field caused by the island structure of the films.

XVI.1. Introduction

It is now possible to make high quality, single–crystalline metallic superlattices with monolayer precision [XVI.16] by molecular beam epitaxy (MBE). Meanwhile, there has developed both theoretical and experimental controversy involving ultrathin Fe films, over questions such as the value of the magnetic moment at the Fe surface [XVI.2], the magnetic anisotropy [XVI.3], and reported observations of a reduced Curie temperature [XVI.4–XVI.7].

In this paper, we present a detailed study of the first Fe(110)/Ag(111) superlattices fabricated by MBE. Analysis by Mössbauer spectroscopy provides direct evidence of a magnetic coupling between neighboring Fe layers across an Ag barrier. Our thinnest Fe component superlattice films exhibit a linear temperature dependence of the hyperfine field. Although this behavior could be mistaken for a two–dimensional effect, we show that it is a result of superparamagnetism.

The superlattices studied were of the form $(Fe_x Ag_y)_{30}$ where x refers to the number of monolayers (ML) of Fe (1 ML=2.0 Å) and y refers to the number of

ML of Ag (1 ML=2.4 Å). All superlattices consisted of 30 repeats. Three series of superlattices were made with Fe thicknesses of x =2, 5 and 8 ML. Each of these series had varying Ag thicknesses of y =5, 12 and 20 ML.

XVI.2. Experimental

The superlattices were produced in a PHI model 430B system which includes *in situ* analysis by reflection high energy electron diffraction (RHEED), a residual gas analyzer, and a quartz crystal oscillator to monitor deposition rates. Ag and Fe (enriched to a concentration of 35% Fe^{57}) were deposited from modified EPI effusion cells onto a synthetic Fe–free single crystal mica substrate. The pressure during film growth was typically better than 2×10^{-9} torr. It was shown in a previous paper [XVI.8] that Ag(111) grows epitaxially on mica when the substrate temperature is held at 180° C, and that Fe(110) will grow epitaxially on Ag(111) with only a single orientation as determined by RHEED. The high quality of the RHEED patterns obtained after deposition of both the first Ag and the first Fe layer are maintained with little degradation even after 100 repetitions of each component. This indicates that the superlattice maintains its flat, single–crystalline nature from the first repetition to the last. Cross-sectional transmission electron microscopy (TEM) was also performed on some of these superlattices and it showed that the respective Fe and Ag bilayers were flat and continuous. Further details of film characterization and growth procedures can be found elsewhere [XVI.8].

XVI.3. The 5ML and 8ML Fe Bilayer Superlattices

Transmission Mössbauer spectroscopy was used to analyze these samples at temperatures from 4.2 K to 300 K. Fig. XVI.1 shows the spectra at 4.2 K for all six films and Fig. XVI.2 shows the same films at room temperature. It is apparent that all these films are magnetically ordered at all the measurement temperatures. There is no evidence of a superparamagnetic component (central broad spectral feature), even at room temperature. This effectively means that the Fe bilayers are continuous on an atomic scale, and do not consist of separated islands. This contrasts with other work, where superparamagnetism was seen below 8 ML of Fe [XVI.9,XVI.10].

The linewidths of all of the spectra are less than 0.35 mm/sec. This supports the view that all of the superlattice bilayers are flat and continuous with all the Fe occupying sites that are essentially the same. The relative intensity of the lines for each spectrum is approximately 3:4:1, indicating that the magnetization of each Fe bilayer is in–plane.

Fig. XVI.3 shows a plot of the mangetic hyperfine field (H) versus temperature for the three 8 ML Fe superlattices with Ag=5, 12 and 20 ML. Fig. XVI.4 shows a similar plot for the three 5 ML Fe superlattices. All the curves are well fitted to the equation

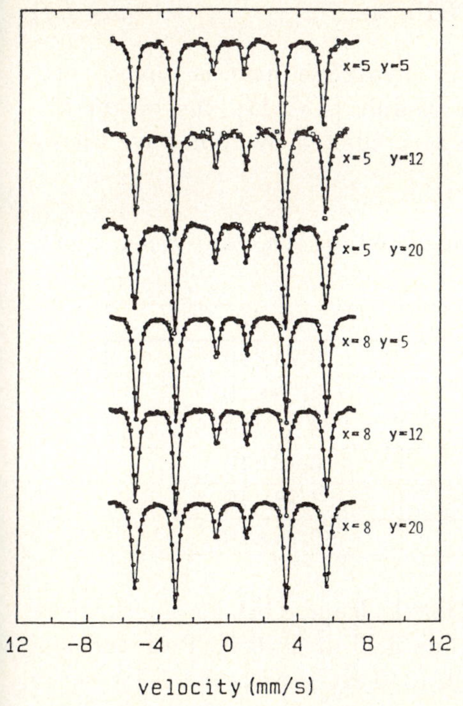

Figure XVI.1. 5 and 8 ML Fe superlattice Mössbauer spectra at 4.2 K.

Figure XVI.2. 5 and 8 ML superlattice Mössbauer spectra at room temperature.

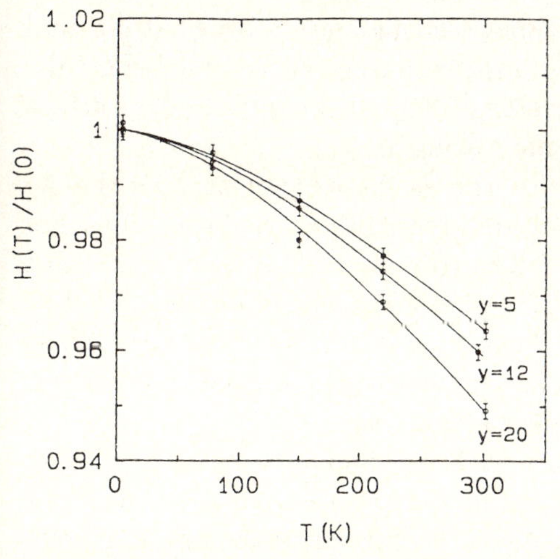

Figure XVI.3. Magnetic hyperfine field (H_o) versus temperature (T) for three 8 ML Fe superlattices.

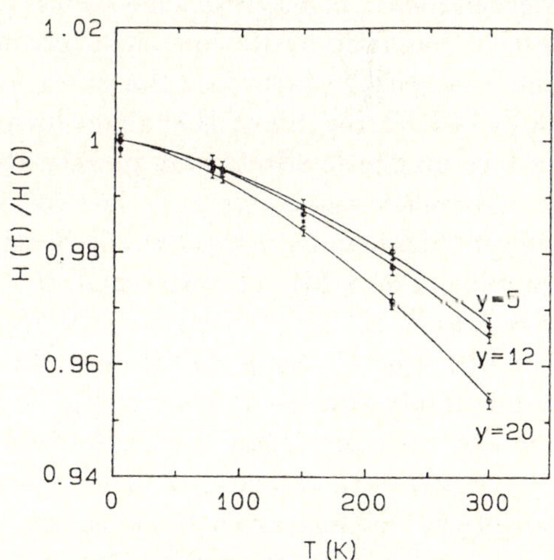

Figure XVI.4. H_o versus T for the three 5 ML Fe superlattices.

141

$$H(T) = H_\circ \left(1 - BT^{3/2}\right) . \qquad\qquad (XVI.1)$$

Since it has been well established that the hyperfine field is closely proportional to the local magnetization in the sample [XVI.11], the magnetization also decreases with a $T^{3/2}$ dependence as predicted by spin–wave theory [XVI.12,XVI.13].

Table XVI.1. Values of $H_\circ$ and B factor measured for the 5 and 8 ML superlattices.

x	y	$H_\circ$	B
5	8	342.5	6.25×10^{-6}
12	8	342.6	6.75×10^{-6}
20	8	342.9	8.81×10^{-6}
5	5	342.9	6.96×10^{-6}
12	5	343.2	7.88×10^{-6}
20	5	343.4	9.75×10^{-6}

Table XVI.1 lists the values of $H_\circ$ and B for the 5 and 8 ML Fe superlattices, where H represents the ground–state hyperfine field at T=0 K. Both series of superlattices have H enhanced over the bulk (340 KG), with the 5 ML series showing the greater enhancement.

The effect of the surface on the magnetic properties of the superlattices is apparent from an examination of the B factor. The value of B for bulk Fe, which is a measure of how easy it is to form collective magnetic excitations (spin–waves), is approximately 5.2×10^{-6} K$^{-3/2}$, determined from a bulk–measurement. It has been shown both theoretically and experimentally that collective magnetic excitations at a ferromagnetic surface are stronger than bulk and that the B factor at the surface is approximately twice the bulk value [XVI.14,XVI.16]. Since B changes by a factor of 2 from the bulk to the surface, surface magnetic effects may be seen by measuring B.

For each respective superlattice series, the B factor increases at the Ag bilayer thickness increases and all B values are greater than the bulk value. For example, the 8 ML Fe series had $B = 6.2 \times 10^{-6}$ K$^{-3/2}$ for $y = 5$ ML and $B = 8.81 \times 10^{-6}$ K$^{-3/2}$ for $y = 20$ ML, and the 5 ML Fe series had $B = 6.96 \times 10^{-6}$ K$^{-3/2}$ for $y = 5$ ML and $B = 9.74 \times 10^{-6}$ K$^{-3/2}$ for $y = 20$ ML. These trends imply that the effects of the surface on the magnetic excitations of the film are more prominent as the Ag thickness increases.

As the data indicate, B varies not only with Ag thickness, but also with Fe thickness. For any given thickness of Ag, the superlattice with the thinner Fe component has the greater B value. Since a greater fraction of the Fe component in these thinner superlattices is at the surface, this enhancement is expected.

The above observations of the properties of the 5 and 8 ML Fe superlattices can be understood in terms of a magnetic interaction between neighboring Fe bilayers separated by Ag. When the Ag bilayer is thin, the interaction between neighboring Fe bilayer surfaces is strong and the result is more bulk–like behavior. As the Ag bilayer thickens, the Fe bilayers become more isolated and they exhibit more surface–like magnetic behavior.

XVI.4. The 2 ML Superlattices

Three superlattices were grown in this series, with Ag bilayer thickness $(y)=5$, 12 and 20 ML. RHEED patterns of these superlattices do not differ significantly from those of the thicker Fe bilayer superlattices discussed earlier, indicating that these films are flat and single–crystalline. DC SQUID measurements were performed on these films. The results, with the applied field both parallel and perpendicular to the film surface, are shown in Fig. XVI.5. There is a strong shape anisotropy between the 2 directions, with a saturation field in the perpendicular direction of approximately 20 kG. This value is approximately equal to the value of the shape anisotropy $4\pi M_s$ (21.5 kG) expected for a flat continuous Fe film. This indicates that, if the Fe bilayers are not continuous but consist of islands, then the island size must be much greater than the Fe thickness and the islands must be flat.

These superlattices were analyzed by transmission Mösbauer spectroscopy form 4.2 K to 450 K. The spectra of $(Fe_2\,Ag_5)_{30}$ at different temperatures are shown in Fig. XVI.6 and were fitted to one sextet plus a single central component. The spectra of the other 2 ML Fe superlattices were similar. The relative intensity of the sextet lines is 3:4:1, indicating that the Fe magnetization is in–plane. Although the intensity of the additional central feature clearly increases as the temperature increases, there is no evidence that the spectrum has totally collapsed even at 500 K, showing that the Curie temperature must be above this value.

The central feature in all these spectra is a superparamagnetic component resulting from the thermal relaxation of the magnetic moment of small Fe islands and is in no way connected with a reduced Curie temperature. This is proven

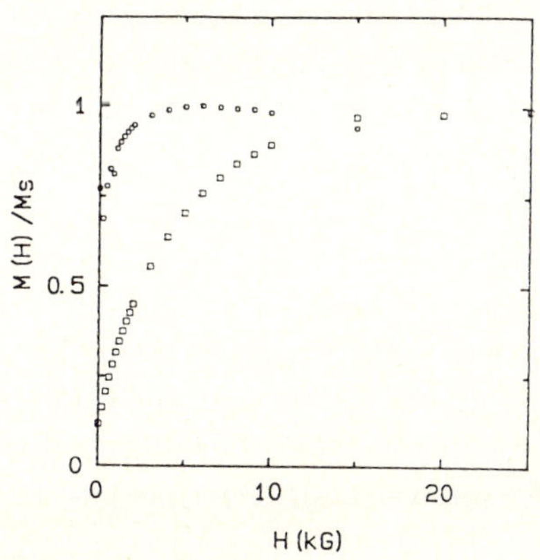

Figure XVI.5. DC SQUID magnetometer plots of M versus applied H for $(Fe_2\,Ag_{12})_{30}$ at $T = 5$ K with H parallel (circles) and perpendicular (squares) to the film.

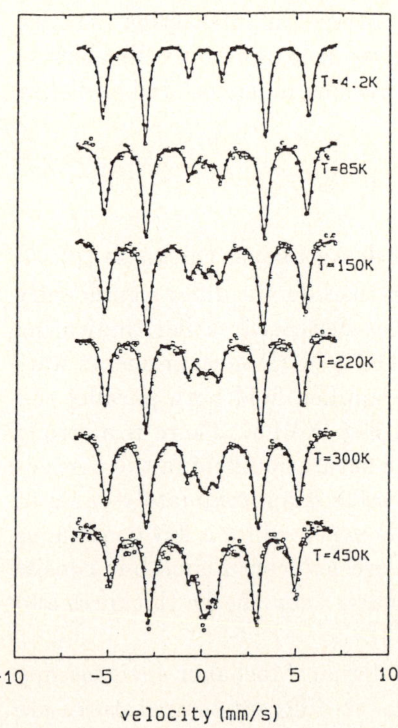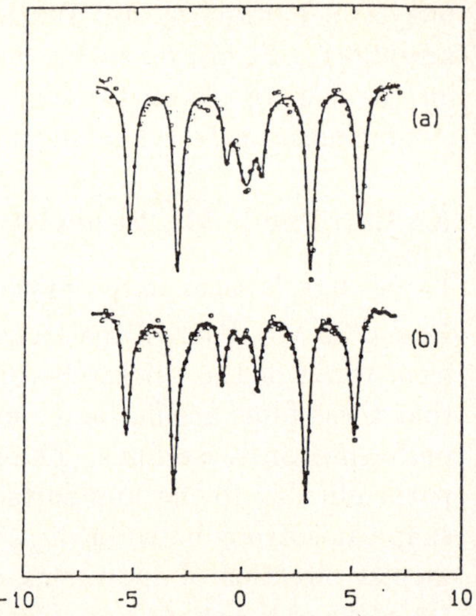

Figure XVI.7. Mössbauer spectra of $(Fe_2\,Ag_{12})_{30}$ at room temperature
a) without external magnetic field,
b) with 5 kG external field parallel to the film.

Figure XVI.6. Mössbauer spectra of $(Fe_2\,Ag_5)_{30}$ at different temperatures.

by application of a small magnetic field in the film plane. Figs. XVI.7a and XVI.7b show Mössbauer spectra at room temperature for $(Fe_2\,Ag_{12})_{30}$, but the spectrum in XVI.7b, with a greatly reduced central feature, was taken with a 5 kG external magnetic field parallel to the film. Without the external field, the central feature makes up approximately 28% of the spectrum's intensity. With the magnetic field, this is reduced to < 3%. The energy barrier supplied by the external field to each of the Fe atoms is about 2.2 μ_B H or 10^{-16} erg, which is much less than the thermal energy $(k_B\,T)$ at room temperature. In other words, the external magnetic field is not strong enough to align the Fe moments individually. Therefore, this central feature cannot be associated with loss of magnetic relaxation of small Fe flat islands, each of which is already ferromagnetically ordered.

Fig. XVI.8 shows that the hyperfine fields of the three superlattices obey a linear temperature dependence rather than the usual $T^{3/2}$ dependence found in 3–D ferromagnets. This linear behavior has frequently been observed for Fe films approaching 1 ML and it has often been attributed to their two–dimensional nature. However, if this were the case, it would be expected that the value of the hyperfine field would decrease more rapidly with temperature as the thickness of

144

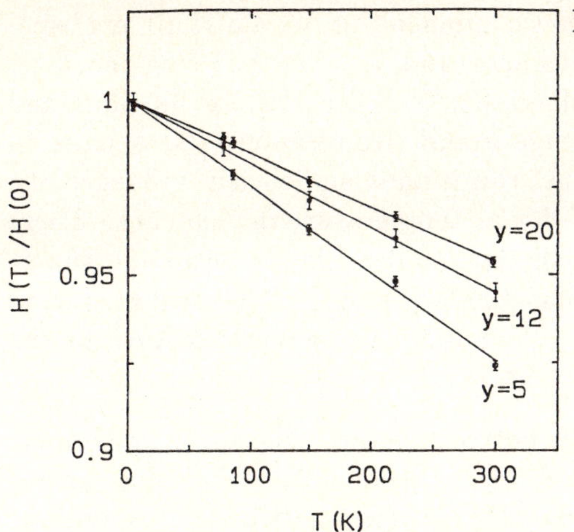

Figure XVI.8. $H_{\circ}$ versus T for the 2 ML Fe superlattices.

Table XVI.2. The temperature dependence of the hyperfine field is fit to
$$H(T) = H_{\circ}(1 - \alpha T).$$

| | $(\text{Ag}_{20}\text{Fe}_2)_{30}$ | $(\text{Ag}_{12}\text{Fe}_2)_{30}$ | $(\text{Ag}_5\text{Fe}_2)_{30}$ | $(\text{Ag}_{20}\text{Fe}_2)_5$ | | |
				a1	a2	a3
$\alpha \times 10^4\,(\text{K}^{-1})$	1.56	1.82	2.48	5.69	3.84 2.60	2.08
The average size of the islands $\langle \ell \rangle$ (Å)	158	146	125	83	100 122	137
The relative weight of the central peak at 295 K	0.23	0.28	0.21	0.63	0.43 0.38	0.39

the Ag component increased. This trend is expected, as we have seen earlier for the 5 and 8 ML superlattices because as the Ag bilayer thickness increases, the interaction between neighboring Fe bilayers weakens, making it easier to excite the Fe magnetic moments [XVI.17]. In fact, as Fig. XVI.8 and Table 2 indicate, the actual trend is just the opposite: as the Ag bilayer becomes thicker, the hyperfine field decreases rapidly.

In order to demonstrate how this linear temperature dependence could result from an island structure, we have developed a simple model relating the slope of the hyperfine field with island size. The details of this model can be found elsewhere [XVI.17]. The key result is that the model yields a quasi–linear temperature dependence:

$$\langle u \rangle = 1 - \frac{k_s T}{2KSd[1 - d_{\circ}/(2d)]} \qquad (XVI.2)$$

145

where $\langle u \rangle$ is the average island magnetic moment, $k = 4.5 \times 10$ erg/cm^3, $S =$island area, $d =$island height or thickness, and $d_0 = 71$ Å. Using the slope values from Fig. XVI.8, an average island size is $\langle 1 \rangle = \langle S \rangle$ for all films and is listed in Table 2. Although the average island size of about 100 Å is much greater than the film's thickness of 4 Å, the islands sufficiently influence the thermal excitations to produce a quasi–linear temperature dependence. There is obviously a range of island sizes, with the smaller islands becoming super-paramagnetic at a lower temperature than the larger ones. It is estimated from our model that, at room temperature, the superparamagnetic feature in the Mössbauer spectra is due to islands 60 Å in size.

In order to test this model, another heterostructure $(Fe_2 Ag_{20})_5$ was made with a reduced substrate temperature of 75° C (versus 85° C). From various models of epitaxial growth, it is expected that a lower substrate temperature promotes the growth of smaller islands. Thus, if our model is correct, this film should exhibit a larger superparamagnetic feature and a hyperfine field with a larger slope than the films grown at higher temperature. Also it is reasonable to expect that these islands may coalesce upon annealing, decreasing both the superparamagnetic component and the slope of the hyperfine field. After taking the Mössbauer spectra of this heterostructure, the film was annealed and remeasured after each anneal. Two anneals were performed in Ar gas for one

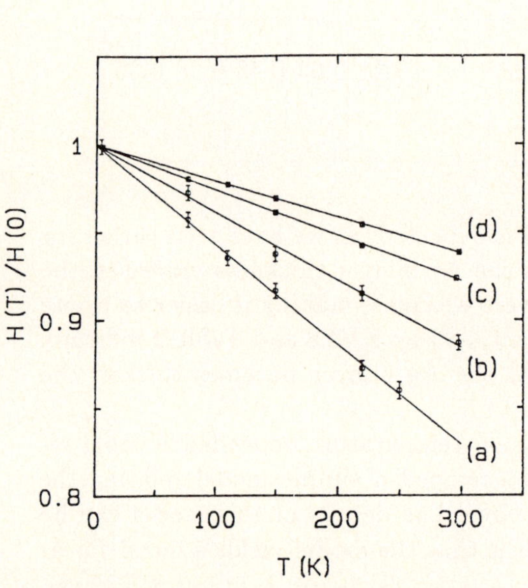

Figure XVI.9. H_0 versus T for the $(Fe_2 Ag_{20})_5$ heterostructure after various phases of anneal.

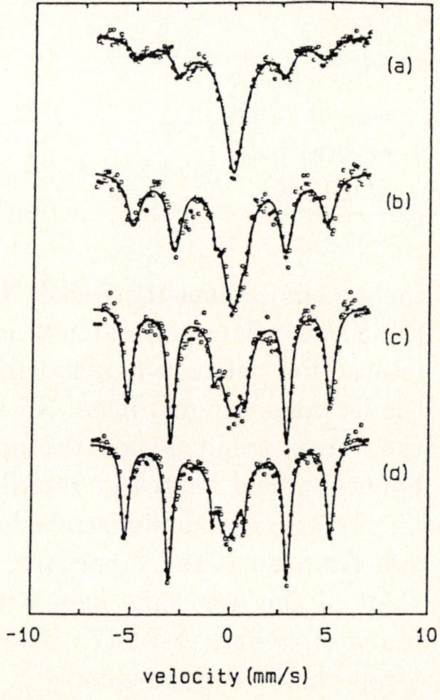

Figure XVI.10. Mössbauer spectra for the $(Fe_2 Ag_{20})_5$ heterostructure at room temperature after various phases of anneal.

hour each at 200 (a1) and 250° C (a2). A final anneal in vacuum for one hour at 350° C was performed (a3). These results are given in Table 2. Figures XVI.9 and XVI.10 show the temperature dependence of the hyperfine field and room temperature Mössbauer spectra, respectively, after each stage of the anneal. The data clearly support our model. After each stage of the anneal, both the slope and the central feature decreases, indicating that the islands have coalesced.

XVI.5. Conclusions

High quality, well–characterized Fe(110)/Ag(111) superlattices have been fabricated for the first time. A magnetic coupling between Fe bilayers was observed for the 5 and 8 ML superlattices. The 2 ML superlattices exhibited a linear temperature dependence which was not a two–dimensional effect, but is a consequence of the varying island structure of these films.

This work has been supported in part by NSF grant DMR–8722352.

References

XVI.1 I. K. Schuller and C. M. Falco, in *Microstructure Science and Engineering/VLSI*, edited by N. G. Einspruch (Academic, New York, 1982).

XVI.2 C. L. Fu, A. J. Freeman and T. Oguchi, Phys. Rev. Lett. **54**, 2700 (1985); C. L. Fu and A. J. Freeman, J. Magn. Magn. Mat. **69**, L1 (1987).

XVI.3 J. G. Gay and R. Richter, Phys. Rev. Lett. **56**, 2728 (1986); J. G. Gay and R. Richter, J. Appl. Phys. **61**, 3362 (1987).

XVI.4 B. T. Jonker, K. H. Walker, E. Kister, G. A. Prinz, and C. Carbone, Phys. Rev. Lett. **57**, 143 (1986); C. Rau, C. Schneider, G. Xing, and K. Jamison, Phys. Rev. Lett. **57**, 3221 (1986).

XVI.5 W. A. A. Macedo and W. Keune, Phys. Rev. Lett. **61**, 475 (1988).

XVI.6 M. Przybylski and U. Gradmann, Phys. Rev. Lett. **59**, 1152 (1987).

XVI.7 M. Przybylski and U. Gradmann, to be published in J. de Physique.

XVI.8 C. J. Gutierrcz, S. H. Mayer and J. C. Walker, J. Magn. Magn. Mat. (accepted for publication).

XVI.9 G. Bayreuther and G. Lugert, J. Magn. Magn. Mat. **35**, 50 (1983).

XVI.10 F. A. Volkening, B. T. Jonker, J. J. Krebs, N. C. Koon, and G. A. Prinz, J. Appl. Phys. **63**, 3869 (1988).

XVI.11 I. Vincze and J. Kollar, Phys. Rev. B **6**, 1066 (1972).

XVI.12 D. L. Mills and A. A. Maradudin, J. Phys. Chem. Solids **28**, 1855 (1967).

XVI.13 J. Mathon, Phys. Rev. B **24**, 6588 (1981).

XVI.14 G. T. Rado, Bull. Am. Phys. Soc. **2**, 127 (1957).

XVI.15 J. Korecki and U. Gradmann, Eur. Lett. **2**, 651 (1986).

XVI.16 J. C. Walker, R. Droste, G. Stern, and J. Tyson, J. Appl. Phys. **55**, 2500 (1984).

XVI.17 Z. Q. Qiu, S. H. Mayer, C. J. Gutierrez, H. Tang, and J. C. Walker, Phys. Rev. Lett. (submitted).

Part 4

Statistical Mechanics of Low-Dimensional Magnetism

XVII. Static and Dynamic Critical Phenomena at Surfaces

S. Dietrich

Fachbereich Physik, Bergische Universität Wuppertal,
Postfach 10 10 27, D-5600 Wuppertal 1, Fed. Rep. of Germany

This paper gives a brief account of the effects of surfaces on critical phenomena. It contains the discussion of both static properties like the phase diagram, surface critical exponents as well as scaling, and of the dynamic behavior of various correlation functions. The analytic field–theoretic results agree well with those obtained from numerical simulations and experiments. Recent extensions of this field of research are mentioned.

XVII.1. Introduction

Critical phenomena exhibit universal properties which depend only on the space dimension d and on internal symmetries $O(n)$. Every sample is bounded by surfaces where these essential ingredients are modified. In a $(d-1)$–dimensional layer near the surface the structural and thermal properties of the systems deviate from their corresponding bulk ones. Close to the surface also the $O(n)$ symmetry of the ordering degrees of freedom may be reduced. For that reason one can expect that the local critical behavior at surfaces exhibits rich phenomena and allows one to gain a deeper insight into the structure of critical phenomena.

The thickness of the surfacelike layer is proportional to the bulk correlation length $\xi \sim \tau^{-\nu}$, $\nu \simeq 0.63$, which diverges upon approaching the critical point T_c, i.e., for $\tau = |(T-T_c)/T_c| \to 0$. Thus this layer is part of the critical bulk system and its thickness can grow without limits. This gives rise to an interesting crossover between surface and bulk properties as function of the distance z from the surface.

Due to universality the subsequent theoretical considerations apply to any second–order bulk phase transition. In order to be concrete one can, as it is done below, associate the order parameter ϕ with the magnetization M of either an Ising $(n=1)$, XY $(n=2)$, or Heisenberg $(n=3)$ ferromagnet.

Since two detailed review articles about this subject have already been published [XVII.1,XVII.2], here I focus only on the most important facts and on some of the recent developments in this field of research.

Springer Proceedings in Physics, Vol. 50   **Magnetic Properties of Low-Dimensional Systems II**
Editors: L.M. Falicov · F. Mejía-Lira · J.L. Morán-López © Springer-Verlag Berlin, Heidelberg 1990

XVII.2. Static Critical Behavior

Consider the spontaneous magnetization in a semi–infinite Ising model confined
to the halfspace $V_+ = \{\vec{x} = (\vec{r}, z) \in \mathbf{R}^d | z \geq 0\}$. In the topmost layer $z = 0$ the
nearest–neighbor coupling is taken to be $J_1 > 0$, otherwise it is $J > 0$. Due to
the broken translational invariance $M = M(z, \tau)$ depends on z and approaches
its bulk value $M(\infty, \tau)$ according to $M(z \to \infty, \tau) - M(\infty, \tau) \sim exp(-z/\xi)$.
Since there are missing bonds the effective field is smaller at the surface than
in the bulk as long as $J_1/J \leq 1.50$ [XVII.3]. Consequently the spontaneous
magnetization decreases upon approaching the surface. Accordingly as function
of the temperature the surface magnetization vanishes more rapidly than in the
bulk:

$$M(0, \tau) \sim \tau^{\beta_1}, \ M(\infty, \tau) \sim \tau^{\beta}. \qquad (XVII.1)$$

For $d = 3$ $\beta \simeq 0.32$ is the critical exponent for the *bulk* order parameter whereas
$\beta_1 = 0.80 > \beta$ is the corresponding critical *surface* exponent [XVII.2]. (Note
that β_1 differs completely from $\beta(d = 2) = 1/8$.) For this so–called *ordinary
transition*, at which the surface orders at the same temperature T_c as the bulk
does, $\beta_1(d = 3)$ is larger than $\beta(d = 3)$. If, however, J_1 is more than about
50% stronger than J, the effect of the missing bonds is overcompensated and
the surface undergoes a phase transition at a temperature $T_s > T_c$. This ordering for $T_c < T < T_s$ is confined to a surface layer of thickness $\xi < \infty$. This
so–called *surface transition* is characterized by the critical exponents of a $(d-1)$–
dimensional bulk system. A further lowering of the temperature leads to the
so–called *extraordinary transition*, where at T_c the bulk orders in the presence of
an already ordered surface. At the surface the latter transition is accompanied
by only weak thermal singularities. For $J_1 = J_{1c} \simeq 1.50J$ the system exhibits a
multicritical surface transition called *special transition*, which separates the ordinary from the extraordinary transition. It turns out that in this case the surface
magnetization vanishes more slowly than in the bulk, $\beta_1^{sp} \simeq 0.24 < \beta$, so that
$M(z, \tau)$ increases upon approaching the surface $\sim z^{(\beta_1 - \beta)/\nu}$ [XVII.2,XVII.3].

Close to criticality the order parameter $M(z, \tau)$ takes on a scaling form:

$$M(z, \tau) = \tau^{\beta} f(z/\xi), \qquad (XVII.2)$$

which is valid for $z \sim a$, where a is the lattice constant. The scaling function
$f(x)$, which is universal up to an amplitude, behaves for large x as $f(x \to
\infty) - f(\infty) \sim e^{-x}$ and is singular for small x as $f(x \to 0) \sim x^{(\beta - \beta_1)/\nu}$. Therefore,
there is a smooth and universal crossover between the surface behavior $\sim \tau^{\beta_1}$
for $z \ll \xi$ and the bulk behavior $\sim \tau^{\beta}$ for $z \gg \xi$.

The semi–infinite geometry requires to consider several distinct susceptibilities:

$$\chi_1 = \frac{\partial M(z = 0)}{\partial H} \sim \tau^{\gamma_1} \qquad (XVII.3)$$

describes the response of the surface magnetization on a variation of the bulk magnetic field H and

$$\chi_{11} = \frac{\partial M(z=0)}{\partial H_1} \sim \tau^{-\gamma_{11}} \qquad (XVII.4)$$

is the corresponding response on a surface field H_1. Eqs. (XVII.3), (XVII.4) hold not only for $z = 0$ but for any $z \ll \xi$. The field–theoretic renormalization group yields $\gamma_1 \simeq 0.77$ and $\gamma_{11} \simeq -0.32$ [XVII.2]. (Here and in the following I quote only the values of the critical exponents for the Ising universality class at the ordinary transition; those for the XY and the Heisenberg class are also known analytically up to second order in $\epsilon = 4 - d$). It should be emphasized that γ_{11} is negative. This means that χ_{11} remains finite at $T = T_c$ but attains its maximum value in a cusplike singularity. This has to be compared with $\gamma \simeq 1.24$ in the bulk.

For experimental investigations of the structural changes at phase transitions the two–point correlation function $G = \langle M(\vec{x})M(\vec{x}') \rangle$ is of particular importance. Its Fourier transform determines the cross–section of scattered X–rays and neutrons (see below). In the bulk and at T_c G depends on $x = |\vec{x} - \vec{x}'|$. For large separations it decays asymptotically $\sim x^{-(d-2+\eta)}$ where $\eta \simeq 0.03$ is known as Fishers's exponent. At T_c also in the semi–infinite geometry G displays a power–law decay. Here, however, it is a function of three variables: z, z', and $r = |\vec{r} - \vec{r}'|$. One finds

$$G(z, z', r) \sim x^{-(d-2+\eta_{\parallel,\perp})} \qquad (XVII.5)$$

with an exponent $\eta_\parallel$ or $\eta_\perp$ depending on whether x increases parallel to the surface (z, z', fixed, $r \to \infty$) or in any other direction (z' fixed, $z \to \infty$, $r \to \infty$ or not). Only if all three distances z, z' and r become large simultaneously one recovers the bulk exponent η. The values of the exponents are $\eta_\parallel \simeq 1.48$ and $\eta_\perp \simeq 0.76$, respectively [XVII.2]. Note that $\eta_\parallel$ is about fifty times larger than the small Fisher exponent in the bulk.

There are many additional surface exponents which describe the large variety of the local critical behavior at surfaces, e.g. concerning the singular H– and H_1–dependence of various surface quantities [XVII.2]. However, they as well as those mentioned above are not independent from each other. The field–theoretic renormalization group allows one to derive —beyond perturbation theory— scaling laws between these exponents as well as the scaling behavior of correlation functions like in Eq. (XVII.2), e.g.:

$$\gamma_1 = \nu(2 - \eta_\perp), \; \gamma_{11} = \nu(1 - \eta_\parallel)$$

$$2\gamma_1 - \gamma_{11} = \gamma + \nu, \; 2\eta_\perp = \eta + \eta_\parallel \qquad (XVII.6)$$

$$\beta_1 = \frac{\nu}{2}(d - 2 + \eta_\parallel), \; etc.$$

As indicated by Eq. (XVII.6) there is one and only one independent surface critical exponent for the ordinary transition. All the other exponents follow from the knowledge of two independent bulk exponents. Since the special transition is a multicritical point, it is characterized by one additional relevant crossover exponent $\varphi \simeq 0.68$, which describes how in the phase diagram the line of surface transition joins the line of extraordinary transitions: $T_s - T_c \sim (J_1 - J_{1c})^{1/\varphi}$. Eq. (XVII.6) is valid also at this special transition.

The results presented above refer to systems with vanishing external fields H and H_1. Now I want to mention at least the particularly interesting case $H = 0$ and $H_1 \neq 0$ [XVII.2]. Due to $H = 0$ the bulk order parameter $M(\infty, \tau)$ is zero for $T > T_c$, but the surface field H_1 induces a nonzero order parameter $M(z, \tau)$ for any finite z. It decays $\sim e^{-z/\xi}$ for $z \geq \xi$ and $\sim z^{-\beta/\nu}$ for $a \ll z \leq \xi$. Thus at T_c one is left with a *mesoscopic* interface profile $\sim z^{-0.51}$ whose zeroth moment does no longer exist: the excess quantity $M_s = \int_0^\infty dz\, M(z, \tau)$ diverges $\sim \tau^{\beta_s}$ with $\beta_s = \beta - \nu \simeq -0.31$. For magnets this situation prevails right at the extraordinary transition or it may be induced by a thin surface layer of a magnetic material, which has a high value of T_c, coating the semi–infinite system of interest, which has a lower value of T_c.

Whereas for magnets $H \neq 0$ has to be carefully prepared, it is unavoidably supplied at the surface of a fluid [XVII.4]. In this case the order parameter is $\rho(z) - \rho_c$ where ρ is the number density and ρ_c its value at T_c. This phenomenon called *critical adsorption* has attracted a lot of theoretical and experimental attention [XVII.4] and provides a particular interesting testing ground for critical phenomena at interfaces.

XVII.3. Dynamic Critical Behavior

At a critical point not only the spatial correlations diverge proportionally to the correlation length ξ but there is also a critical slowing down according to which the relaxation time at T_c, for a fluctuation with a long wavelength $\sim q^{-1}$, diverges as q^{-z} $(q \to 0, \tau = 0)$ and as $\tau^{-\nu z}$ $(q = 0, \tau \to 0)$, respectively. The dynamic critical exponent z takes on the value 2.01 for a non–conserved (model A) and 3.97 for a conserved (model B) order parameter [XVII.5]. This critical slowing down can be probed experimentally by inelastic neutron scattering at appropriate model systems. The cross–section is proportional to $C(q, \omega)$, which is the Fourier transform of the two–point correlation funtion with respect to space $(\to \vec{q})$ and time $(\to \omega)$. In the bulk $C(q, \omega)$ as function of ω is peaked at $\omega = 0$ and has a shrinking width proportional to the inverse relaxation time (see above). For $q = 0$ the peak value diverges $\sim \tau^{-\sigma_\tau^{(b)}}$ with $\sigma_\tau^{(b)} = \gamma + \nu z \simeq 2.51$ and for $\tau = 0$ it diverges $\sim q^{-\sigma_q^{(b)}}$ with $\sigma_q^{(b)} = 2 + z - \eta \simeq 3.98$. At T_c and for $q = 0$, $C(q = 0, \omega \to 0)$ it diverges as function of ω as $\omega^{-\sigma_\omega^{(b)}}$ with $\sigma_\omega^{(b)} = 1 + \gamma/(\nu z) \simeq 1.99$. These and the following values for the critical exponents correspond to model A.

By applying field–theoretic techniques the effects of surfaces on these dynamic critical phenomena have been analyzed systematically for model A and B [XVII.6]. The main results can be illustrated by discussing the frequency dependence of the two–point correlation function parallel to the surface, which determines the inelastic cross–section of totally reflected neutrons [XVII.7]. First, one finds that there is no separate, independent dynamic critical exponent associated with the surface. Therefore the linewidth of the scattering function follows the same power law as in the bulk (see above). However, the various divergences at the peak value are significantly weaker than in the bulk: $\sigma_r^{(s)} = \gamma_{11} + \nu z \simeq 0.93$, $\sigma_p^{(s)} = 1 + z - \eta \simeq 1.53$, and $\sigma_\omega^{(s)} = 1 + \gamma_{11}/(\nu z) \simeq 0.76$; $\vec{p}$ denotes the parallel momentum transfer instead of $\vec{q}$ in the bulk. Note that $\int_{-\infty}^{\infty} d\omega\, C(p,\omega) = G(p)$ yields the static structure factor, which exhibits only cusplike singularities (see above).

The study of inelastically scattered neutrons under the conditions of total reflection is a very demanding proposition. Therefore it is very desirable to have additional options for studying the local dynamic critical behavior at interfaces. The possibility is offered by the broadening of the linewidth of Mössbauer (or NMR) spectra, which is determined by the frequency zero mode of the auto–correlation function [XVII.8,XVII.9],

$$A(z,\tau) = \int_{-\infty}^{\infty} dt \langle M(\vec{x},t) M(\vec{x},0) \rangle; \qquad (XVII.7)$$

t denotes the time variable. This method can be made surface specific by placing Mössbauer active nuclei at a specific depth below the surface using epitaxial techniques. Another possibility is to study the linewidth of only those Mössbauer quanta which leave a sample, being homogenously filled with active nuclei, below their corresponding angle of total reflection. These quanta stem from the first few atomic layers of the semi–infinite system and thus they determine $A(z = 0,\tau)$.

By applying field–theoretic methods one finds that in the bulk the auto–correlation function diverges upon approaching T_c whereas in the surface layer $z \ll \xi$ it reaches a maximum value in a cusplike singularity [XVII.10]:

$$A(z = \infty, \tau) \sim \tau^{-\sigma_a^{(b)}}, \quad A(z = 0, \tau) \sim \tau^{-\sigma_a^{(s)}} \qquad (XVII.8)$$

with $\sigma_a^{(b)} = \nu z - 2\beta \simeq 0.63 > 0$ and $\sigma_a^{(s)} = \nu z - 2\beta_1 \simeq -0.33 < 0$ for model A. Similarly, as in Eq. (XVII.2), $A(z,\tau)$ takes on a scaling form whose corresponding universal scaling function mediates smoothly between the surface and bulk behavior. However, the scaling properties of $A(z,\tau)$ are complicated by the fact that the coalescence of the two points $\vec{x}$ and $\vec{x}'$ in the correlation function causes additional singularities which must be cured by an additive renormalization sheme [XVII.10]. The qualitative difference of the broadening of the line width in the bulk, where it diverges, and at the surface, where it reaches a finite value, should yield a valuable test of interfacial dynamics.

The above analysis for the critical dynamics is based on thermal equilibrium. It can be extended in a systematic way to the problem of nonequilibrium phenomena close to T_c and in a semi–infinite geometry [XVII.11]. Consider a magnet at T_c in the presence of a bulk magnet field $H \neq 0$ so that $M(z, t < 0) \neq 0$. At $t = 0$ H is switched off so that for $t \to \infty$ the magnetization relaxes towards its new equilibrium value, which is zero. Due to the critical slowing down M does not relax exponentially but according to a power law:

$$M(z = \infty, t \to \infty) \sim t^{-\sigma_t^{(b)}}, \quad M(z = 0, t \to \infty) \sim t^{\sigma_t^{(s)}} \qquad (XVII.9)$$

with $\sigma_t^{(b)} = \beta/(\nu z) \simeq 0.25$ and $\sigma_t^{(s)} = \beta_1/(\nu z) \simeq 0.63$. Since here $\xi = \infty$, for *any* fixed value of $z < \infty$ there is a crossover time after which the decay of M is governed by $\sigma_t^{(s)}$. This demonstrates that at T_c the effect of the surface has penetrated into the whole sample! This picture has been confirmed by Monte Carlo simulations [XVII.12].

XVII.4. Conclusion

The analysis of critical phenomena at interfaces has provided a large body of theoretical predictions. Some of them have been subjected to numerical tests which confirm quantitatively the above picture [XVII.1–XVII.3,XVII.12]; for some recent results see Refs. XVII.13 and XVII.14. On the experimental side the standard surface–sensitive techniques are rapidly improving and new ones, like the use of totally reflected X–rays and neutrons, yield particularly detailed structural information [XVII.7,XVII.15]. Most of the experiments have been devoted to study the surface magnetization. Many of them confirmed the theoretical predictions for the value of β_1 [XVII.2], others require additional reasearch [XVII.16]. There seems to be even evidence for the existence of a surface transition, but all details of these phenomena are not yet settled (see e.g. Refs. XVII.17–XVII.19). It is particularly valuable that meanwhile surface critical phenomena have been successfully investigated also at *fluid* interfaces [XVII.20,XVII.21]. The agreement between the values for critical surface exponents found there with the theoretical predictions and those for magnetic systems underscores the validity of the concept of universality for surface critical phenomena.

Finally I want to mention briefly, and with no intention of completeness, some of those theoretical advances which are not covered by the two review articles mentioned in the Introduction [XVII.1,XVII.2]. In conjunction with the $(4 - \epsilon)$–expansion the study of the nonlinear σ model with a free surface yields very accurate values for surface critical exponents of the Heisenberg universality class [XVII.22,XVII.13]. The field–theoretic analysis has been extended also to various other models: tricritical systems [XVII.23,XVII.24], Potts model and percolation [XVII.25], and to systems with a Lifshitz point [XVII.26]. Recently the critical dynamics at surfaces has been studied for model C [XVII.27] and for the Kapitza resistance of ^{4}He within model F [XVII.28]. Finally I want to mention that anomalies at T_c of activated processes at surfaces, known as the Hedvall effect, also reflect those singularities induced by the surface critical phenomena of the substrate [XVII.29,XVII.30]. The case $H > 0$ and $H_1 < 0$ leads to completely new, so–called *wetting phenomena* [XVII.4].

References

XVII.1 K. Binder, in *Phase Transitions and Critical Phenomena*, edited by C. Domb and J. L. Lebowitz (Academic, London, 1983),Vol.8, p.1.

XVII.2 H. W. Diehl, in *Phase Transitions and Critical Phenomena*, edited by C. Domb and J. L. Lebowitz (Academic, London, 1986), Vol.10, p.75.

XVII.3 K. Binder and D. P. Landau, Phys. Rev. Lett. **52**, 318 (1984).

XVII.4 S. Dietrich, in *Phase Transitions and Phenomena*, edited by C. Domb and J. L. Lebowitz (Academic, London, 1988), Vol.12, p.1.

XVII.5 P. C. Hohenberg and B. I. Halperin, Rev. Mod. Phys. **49**, 435 (1977).

XVII.6 S. Dietrich and H. W. Diehl, Z. Phys. B **51**, 343 (1983).

XVII.7 S. Dietrich, J. Magn. Magn. Mat. **54–57**, 658 (1986).

XVII.8 H. Wegener, Z. Phys. **186**, 498 (1965).

XVII.9 E. Bradford and W. Marshall, Proc. Phys. Soc. **27**, 731 (1966).

XVII.10 S. Dietrich and H. Kinzelbach, unpublished, and H. Kinzelbach, Diploma thesis, Universität München (1986).

XVII.11 H. Riecke, S. Dietrich, and H. Wagner, Phys. Rev. Lett. **55**, 3010 (1985).

XVII.12 M Kikuchi and Y. Okabe, Phys. Rev. Lett. **55**, 1220 (1985).

XVII.13 M. P. Nightingale and H. W. Blote, Phys. Rev. Lett. **60**, 1562 (1988).

XVII.14 D. P. Landau, R. Pandey, and K. Binder, *Monte Carlo Study of Surface Critical Behavior in the XY–Model*, preprint (1989).

XVII.15 S. Dietrich, *Phase Transitions at Interfaces*, preprint (1989).

XVII.16 B. H. Dauth, S. F. Alvarado, and M. Campagna, Phys. Rev. Lett. **58**, 2118 (1987).

XVII.17 D. Weller, S. F. Alvarado, W. Gudat, K. Shröder, and M. Campagna, Phys. Rev. Lett. **54**, 1555 (1985).

XVII.18 C. Rau and M. Robert, Phys. Rev. Lett. **58**, 2714 (1987).

XVII.19 D. Weller and S. F. Alvarado, Phys. Rev. B **37**, 9911 (1988).

XVII.20 L. Sigl and W. Fenzl, Phys. Rev. Lett. **57**, 2191 (1986).

XVII.21 W. Fenzl, Habilitation Thesis, Universität München (1989).

XVII.22 H. W. Diehl and A. Nüsser, Phys. Rev. Lett. **56**, 2834 (1986).

XVII.23 H. W. Diehl and E. Einsenriegler, Europhys, Lett. **4**, 709 (1987).

XVII.24 E. Einsenriegler and H. W. Diehl, Phys. Rev. B **37**, 5257 (1988).

XVII.25 H. W. Diehl and P. M. Lam, Z. Phys. B **74**, 395 (1989).

XVII.26 G. Gumbs, Phys. Rev. B **33**, 6500 (1986).

XVII.27 G. M. Xiong and C. D. Dong, Z. Phys. B **74**, 379 (1989).

XVII.28 D. Frank and V. Dohm, Phys. Rev. Lett. **62**, 1864 (1989).

XVII.29 U. Seifert and S. Dietrich, Europhys. Lett **3**, 593 (1987).

XVII.30 U. Seifert and H. Wagner, *Critical Magnetic Field Dependence of Thermally Actived Surface Processes*, preprint (1989).

XVIII. Surface-Induced Disorder and Surface Melting

R. Lipowsky

Sektion Physik der Universität München, Theresienstr. 37,
D-8000 München 2, Fed. Rep. of Germany

XVIII.1. Introduction and Outline

Consider a uniaxial antiferromagnet on a simple cubic lattice in an external magnetic field. The lattice can be divided into two sublattices such that, at low temperature T, all spins within one sublattice point in the same direction. In fact, there are two distinct ordered states, at a given T, which can be transformed into one another by interchanging the two sublattices. With increasing T, the magnet becomes less ordered up to a transition temperature $T = T_\star$ at which it undergoes a phase transition to a disordered state. I will be concerned here with the case where this transition is *first order*. Thus, if we look into the *bulk* of such an antiferromagnet, we will see *discontinuous* change from the ordered to the disordered state.

Now, assume that we look instead at the *surface* of such a magnet. Naively, one would expect to see again a discontinuous change in the order. However, theoretical models typically lead to a rather different behavior since they predict a *continuous* decrease of the surface order for a wide range of microscopic coupling parameters. In these cases, any surface order parameter, M_1, is found to behave as

$$M_1 \sim (T_\star - T)^{\beta_1} \text{ with } \beta_1 > 0. \qquad (XVIII.1)$$

The basic physical mechanism for the continuous surface behavior is as follows. Let us decompose the semi–infinite system into a 2–dimensional surface region and a 3–dimensional bulk domain. If there were no coupling between those two systems, the surface would become disordered at its own transition temperature, $T = T_\star^{(s)}$. If the couplings within the surface are comparable to those within the bulk, mean field theory leads to the rough estimate $T_\star^s \cong 2T_\star/3$. Thus for the temperature interval $T_\star^{(s)} < T < T_\star$, the surface would be disordered while the bulk is ordered. Of course, this picture is too crude since the surface *is* coupled to the bulk. On the one hand, this coupling leads to the residual order within the surface region as given by (XVIII.1) as long as $T < T\star$. On the other hand, it implies that the disordering, which is initiated at the surface, spreads into the interior of the crystal. As a result, a whole layer or film of the (nearly) disordered phase intrudes between the surface and the ordered bulk. This process has been termed surface–induced disorder (SID) [XVIII.1].

Springer Proceedings in Physics, Vol. 50   **Magnetic Properties of Low-Dimensional Systems II**
Editors: L.M. Falicov · F. Mejía-Lira · J.L. Morán-López © Springer-Verlag Berlin, Heidelberg 1990

It is obvious from the picture just described that SID is *not* restricted to first–order transitions in antiferromagnets. In fact, such a behavior was first found theoretically for the (100) surface of the Potts model on a simple cubic lattice [XVIII.2,XVIII.3]. In this case one has $q \geq 3$ ordered states which undergo a first–order transition to the disordered phase. More generally, SID can occur for any first–order phase transition with spontaneously broken symmetry.

One example for SID which has been experimentally studied in some detail is the (100) surface of $Cu_3 Au$. This binary alloy undergoes an order–disorder transition at $T_* = 663\,K$ which is strongly first–order in the bulk. In contrast, some LEED data indicated that the surface order vanishes in a continuous fashion [XVIII.4], which led us to suggest that this system provides an example for SID [XVIII.5]. During the last five years, more experimental data have been obtained by LEED [XVIII.6], spin–polarized LEED [XVIII.7], and X–ray diffraction under total external reflection [XVIII.8], which are all consistent with this view.

So far, I have discussed phase transformations within a crystal which do not affect the translational order of the crystal lattice. Now, let us consider the most drastic change in the translational order of the crystal, i.e., melting. The idea that melting often starts at the crystal surface has a long history [XVIII.9]. However, it has been realized only recently that the process of surface melting can lead to surface critical behavior as in (XVIII.1) [XVIII.1]. Much of the recent experimental work on surface melting has focused on Pb, for which it was observed by ion scattering [XVIII.10,XVIII.11] and LEED [XVIII.12]. In addition, surface melting has been recently studied for Ar [XVIII.13], O_2 [XVIII.14], methane [XVIII.15], Ne [XVIII.16] and Ge [XVIII.17].

In the remainder of this paper, I will give a brief review on surface–induced disorder and surface melting, which emphasizes the basic theoretical concepts. First, the free energy of the disordered surface layer is discussed in Sec. 2. This free energy contains the direct interaction of the two interfaces bounding the layer. The form of this interaction depends on the intermolecular forces (Sec. 3). As a consequence, the critical exponents β_1 which govern the surface order parameters as in (XVIII.1) are, in general, non–universal and depend on microscopic parameters (Sec. 4). Finally, I will give a brief outlook on open questions (Sec. 5).

XVIII.2. Free Energy of the Disordered Surface Layer

Consider a disordered surface layer of thickness ℓ, and let us estimate its free energy per unit area, $V(\ell)$. First of all, one must note that the interface between the crystal and the vapor (or 'vacuum') surrounding it now consists of two interfaces: (i) the (vd) interface between the disordered surface phase and the vapor, and (ii) the (do) interface between the disordered surface layer and the ordered bulk state of the crystal. Thus, the free energy $V(\ell)$ may also be regarded as the *effective interaction* between two interfaces.

The leading term of $V(\ell)$ for large ℓ arises from the fact that the disordered phase is metastable rather than stable for $T < T_*$. Close to T_*, the bulk free energy per unit volume, f_{vo}, of the ordered crystalline phase, (which coexists with the vapor) behaves as $f_{vo} \approx f^* - S_{vo}^*(T-T_*)$. Likewise, the bulk free energy, $\hat{f}_d$, of the metastable disordered phase is given by $\hat{f}_d \approx f^* - S_d^*(T-T_*)$. Then the difference in the bulk free energies f_{vo} and $\hat{f}_d$ leads to $V(\ell) \approx \Delta S(T - T_*)$ with $\Delta S \equiv S_d^* - S_{vo}^*$ for a layer of thickness ℓ.

At $T = T_*$, the disordered phase coexists with the ordered phase (and with the vapor), and the leading term of $V(\ell)$ is given by the sum, $\Sigma_{vd} + \Sigma_{do}$, of the interfacial tensions of the two interfaces. The next–to–leading terms, which reflect the nature of the underlying intermolecular forces, will be referred to as the *direct interaction*, $V_{DI}(\ell)$, between the (vd) and the (do) interfaces. Thus, the free energy per unit area of the disordered surface layer has the generic form

$$V(\ell) \approx \Delta S(T_* - T)\ell + \Sigma_{vd} + \Sigma_{do} + V_{DI}(\ell) \qquad (XVIII.2)$$

for large ℓ with $V_{DI}(\ell = \infty) = 0$.

In mean–field theory, the equilibrium value of ℓ follows from [XVIII.1] $\partial V/\partial \ell = 0$ or

$$\Delta S(T_* - T) = -\partial V_{DI}(\ell)/\partial \ell. \qquad (XVIII.3)$$

Two cases must be distinguished depending on the sign of $V_{DI}(\ell)$ for large ℓ. First, assume that $V_{DI}(\ell)$ and, thus, $-\partial V_{DI}/\partial \ell$ is positive for large ℓ. In this case, (XVIII.3) implies that the layer becomes thicker and thicker as T_* is approached from below [XVIII.18], and the disordered surface layer becomes more and more similar to the disordered bulk phase. This is the process of *complete* surface–induced disorder (or surface melting) with $\Sigma_{vo} = \Sigma_{vd} + \Sigma_{do}$ at $T = T_*$. On the other hand, one may have $V_{DI}(\ell) < 0$ and thus $-\partial V_{DI}/\partial \ell < 0$ for large ℓ. In the latter case, either no disordered layer appers in the surface region or the thickness of this layer remains finite as T_* is attained [XVIII.18]. This is the case of *incomplete* surface–induced disorder (or surface melting) with $\Sigma_{vo} < \Sigma_{vd} + \Sigma_{do}$ at $T = T_*$.

In a fluid context, the two relations $\Sigma_{\alpha\gamma} = \Sigma_{\alpha\beta} + \Sigma_{\beta\gamma}$ and $\Sigma_{\alpha\gamma} < \Sigma_{\alpha\beta} + \Sigma_{\beta\gamma}$ for the interfacial tensions of three coexisting phases α, β and γ are well–known and correspond to complete and incomplete wetting, respectively [XVIII.19]. Thus, surface–induced disorder (or surface melting) and wetting belong to the same class of interfacial phase transitions [XVIII.20,XVIII.21,XVIII.22]. However, in contrast to fluid systems, the interfacial tensions Σ_{vc}, Σ_{vd}, and Σ_{do} for the systems considered here depend on the *orientation* of the crystal surface. Therefore, the relation $\Sigma_{vo} = \Sigma_{vd} + \Sigma_{do}$ may hold for some surface orientations, but not for others, and a macroscopic crystal can exhibit both ordered and disordered facets (or rounded portions of its surface) at the same time. This *anisotropy* of the surface behavior has been observed for the surface melting of Pb [XVIII.11].

XVIII.3. Direct Interaction of Interfaces Bounding the Surface Layer

As mentioned above, the term $V_{DI}(\ell)$ in (XVIII.2) represents the direct interaction between the two interfaces bounding the surface layer, and arises from the microscopic forces between atoms and molecules. First, let us assume that the short–range part of these forces is strong, and let us ignore possible contributions to $V_{DI}(\ell)$ from their long–range part. Then, $V_{DI}(\ell)$ decays *exponentially* for large ℓ. The corresponding decay length can arise from an *order parameter* (OP) density or from a *non–ordering* (NO) density.

In general, a macroscopic system can be described by several OP and NO densities. By definition, all OP densities vanish and all NO densities have a finite value when the system is in disordered high-T state. For example, the Potts model on a simple cubic lattice with $q = 3, 4, \dots$ states is described by $(q - 1)$ OP densities. Likewise, an antiferromagnet on a simple cubic lattice is characterized by one OP and one NO density which are given by the difference and the sum of the sublattice magnetizations, respectively.

For a system governed by short–range forces, all densities decay exponentially $\sim \exp(-z/a)$ with the distance z from an interface but the size of the decay length, a, depends on the density considered. Let ξ_d denote the *largest* decay length of all OP densities in the disordered phase, and let κ_d be the largest decay length of all NO densities [XVIII.23]. Then, two cases must be distinguished. For $\xi_d/2 > \kappa_d$, the OP density governed by ξ_d dominates and the direct interaction decays as [XVIII.1,XVIII.24]

$$V_{DI} \approx c_1 \, \Sigma_{do} \, exp[-2\ell/\xi_d] \qquad (\xi_d/2 > \kappa_d) \qquad\qquad (XVIII.4)$$

for large ℓ where $c_1 \sim O(1)$ is a numerical coefficient.

It is interesting to note that the decay length for $V_{DI}(\ell)$ is $\xi_d/2$ rather than ξ_d. This property is a consequence of the fact that the systems considered here have several ordered phases for $T < T_*$ (which are related by a discrete symmetry) and, thus, have a vanishing field conjugate to the OP densities.

On the other hand, for $\kappa_d > \xi_d/2$, the NO density governed by κ_d dominates and the direct interaction has the asymptotic behavior [XVIII.25]

$$V_{DI}(\ell) \approx \bar{c}_1 \, \Sigma_{do} \, exp[-\ell/\kappa_d] \qquad (\kappa_d > \xi_d/2) \qquad\qquad (XVIII.5)$$

for large ℓ. Usually, the numerical coefficient $\bar{c}_1$ is again $\sim O(1)$ but, in some cases, $\bar{c}_1$ is exceptionally small. Then, one must include the next–order term which is given by $exp[-2\ell/\kappa_d]$, $exp[-2\ell/\xi_d]$ or $exp[-\ell/\kappa'_d]$ where κ'_d is the second largest decay length of the NO densities.

Next, let us consider the contribution to $V_{DI}(\ell)$ which arises from the long–range part of the intermolecular forces. These contributions lead to a direct interaction,

$$V_{DI}(\ell) \approx -W/\ell^{r} \quad \text{for large } \ell, \qquad (XVIII.6)$$

which decays as a power law. Indeed, all atoms or molecules interact via long–range van der Waals forces. These forces give a direct interaction $V_{DI}(\ell) \sim 1/\ell^{r}$ for large ℓ with $r = 2$ and $r = 3$ for nonretarded and retarded forces respectively in 3–dimensional systems [XVIII.26].

XVIII.4. Non–Universal Behavior of Surface Order Parameters

In a semi–infinite system bounded by a planar surface, the OP densities depend on the distance z from the surface $z = 0$. Now, let $m^{(j)}(z)$ with $j = 1, 2, \ldots$ denote the different OP profiles. In order to simplify the discussion, I will assume in the following that these profiles *decay exponentially* with z: $m^{(j)}(z) \sim exp(-z/\xi_d^{(j)})$ where the decay lengths $\xi_d^{(j)}$ are ordered according to their size, i.e. $\xi_d^{(j)} \geq \xi_d^{(j+1)}$ with $\xi_d^{(1)} \equiv \xi_d$.

Within mean–field theory, the equilibrium value of the layer thickness, ℓ, follows from $(T_{\star} - T) \sim -\partial V_{DI}/\partial \ell$ as in (XVIII.3). Then, the mean–field behavior of the surface OPs, $m^{(j)}(z = 0)$, is given by

$$M_1^{(j)} \equiv m^{(j)}(z = 0) \sim exp[-\ell/\xi_d^{(j)}]. \qquad (XVIII.7)$$

For systems governed by short–range forces, the direct interaction V_{DI} has the form as given by (XVIII.4) or (XVIII.5). First assume that $\xi_d/2 > \kappa_d$, i.e. that the dominant length scale belongs to an OP density. Then (XVIII.4), (XVIII.3), and (XVIII.7) imply the mean–field critical behavior

$$\ell \approx \frac{1}{2}\, \xi_d \, \ln[T_{SC}/(T_{\star} - T)] \qquad (XVIII.8)$$

and

$$M_i^{(j)} \sim (T_{\star} - T)^{\beta_1} \quad \text{with} \quad \beta_1 = \xi_d/2\xi_d^{(j)}. \qquad (XVIII.9)$$

as $T_{\star}$ is approached from below. Thus, the dominant OP density vanishes as $M_1^{(1)} \sim (T_{\star} - T)^{\beta_1}$ with the universal critical exponent $\beta_1 = 1.2$ [XVIII.3].

Next, consider the case for short–range forces with $\kappa_d > \xi_d/2$ where κ_d is again the decay length of the dominant NO density. Then, $V_{DI}(\ell)$ is given by (XVIII.5) which leads to

$$\ell \approx \kappa_d \, \ln[T_{SC}/(T_{\star} - T)] \qquad (XVIII.10)$$

and

$$M^{(j)} \sim (T_\star - T)^{\beta_1} \quad \text{with} \quad \beta_1 = \kappa_{\mathrm{d}}/\xi^{(j)} \qquad (XVIII.11)$$

for small $(T_\star - T)$. Thus, the dominant OP density vanishes at the surface as $M_1^{(1)} \sim (T_\star - T)^{\beta_1}$ with the parameter–dependent critical exponent $\beta_1 = \kappa_d/\xi_d$ [XVIII.25].

In general, the form of the direct interaction $V_{DI}(\ell)$ depends on the surface orientation. Therefore, the critical exponent β_1 as given by (XVIII.11) or (XVIII.13) will also depend on this orientation. This has been shown explicitly for a spin (or lattice gas) model which describes a binary alloy on a fcc lattice. In this case, the system is described by one OP density and one NO density which are given by the Bragg–Williams OP and the composition, respectively [XVIII.27,XVIII.25]. Within mean–field theory, the OP is governed by $\beta_1 = 1/2$ for the (111) and the (110) surfaces of the fcc lattice, and by $\beta_1 \cong 2.2 - 2.8$ for the (100) surface where, in the latter case, the precise value depends on the coupling parameters [XVIII.25].

The critical behavior as given by (XVIII.11) has also been obtained within a phenomenological model for surface melting [XVIII.28]. Quite generally, the atomic density of a semi–infinite crystal can be parametrized by the reciprocal lattice vectors, $\vec{Q}''$, of the 2–dimensional lattice parallel to the surface. Within the model of Ref. XVIII.28, $\tilde{M}(\vec{Q}'' = 0, z)$ is treated as a NO density while all $\tilde{M}(\vec{Q}'', z)$ with $|\vec{Q}''| > 0$ are treated as OP densities [XVIII.29]. One then finds that the surface OPs vanish as in (XVIII.11) with

$$\beta_1(\vec{Q}_2'') > \beta_1(\vec{Q}_1'') \quad \text{for} \quad |\vec{Q}_2''| > |\vec{Q}_1''|. \qquad (XVIII.12)$$

Thus, the OP components with *larger* $|\vec{Q}''|$ are predicted to vanish *faster* as $T_\star$ is approached. This is in qualitative agreement with recent LEED experiments on Pb [XVIII.12], and is expected to hold in more detailed models of surface melting.

Finally, consider the case of long–range forces and $V_{DI}(\ell) \sim 1/\ell^r$ as in (XVIII.6). Such an interaction with $r = 2$ (arising from non–retarded van der Waals forces) should apply to surface melting as soon as the disordered layer gets sufficiently thick [XVIII.21,XVIII.16,XVIII.30]. Then, the disordered surface layer thickens as [XVIII.31]

$$\ell \sim 1/(T_\star - T)^\psi \quad \text{with} \quad \psi = -\beta_s = 1/(1 + r), \qquad (XVIII.13)$$

and the surface OPs vanish as [XVIII.31–33,XVIII.28]

$$M^{(j)} \sim exp[-c^{(j)}/(T_\star - T)^\psi] \qquad (XVIII.14)$$

provided the OP profiles decay exponentially as assumed.

XVIII.5. Summary and Outlook

In summary, the surface of a crystal may initiate the disordering process near a first–order phase transition well below the transition temperature $T = T_\star$. Then, the surface exhibits critical behavior and the surface OPs vanish as in (XVIII.1) with characteristic critical exponents β_1. In general, these exponents are non–universal and reflect the relative size of microscopic length scales [see (XVIII.9), (XVIII.11), and (XVIII.12)].

In some systems, the surface favors the ordered state rather than the disordered one. Such surface–induced order [XVIII.1] or surface freezing has been observed experimentally at first–order phase transitions between crystalline phases [XVIII.34]. In these cases, the surface OPs do not vanish, and the thickening of the ordered surface layer is governed by the direct interactions (XVIII.5) or (XVIII.6) which leads to the growth laws (XVIII.10) or (XVIII.13).

There are several *theoretical* issues which require further study. The most interesting points seem to be: (i) More detailed models for the surface melting. In particular, one would like to include the decay of translational order perpendicular to the surface which has been ignored in Ref. (XVIII.28). This problem can be addressed by density functional theory [XVIII.33,XVIII.35]. (ii) The influence of long–range oscillatory interactions in metals and alloys. (iii) The influence of interfacial fluctuations which I have not addressed in this paper. For the direct interaction as given by (XVIII.6) or (XVIII.7) , interfacial fluctuations are expected to lead to a non–trivial renormalization of the critical behavior [XVIII.24,XVIII.21,XVIII.36]. It is not clear, however, if these fluctuation effects will be visible in experiments, since computer simulations of lattice models have not produced, so far, any evidence for them [XVIII.37,XVIII.38].

Needles to say that more *experimental* work is highly desirable. A very promising tool is X–ray diffraction under total external reflection [XVIII.39] which has already been applied to $Cu_3 Au$ [XVIII.8]. In particular, it would be interesting to look for surface–induced disorder in magnetic materials with first–order phase transitions.

I thank H. Bonzel, U. Breuer, and K. Prince for stimulating collaboration, and Dan Kroll, Hartmut Löwen, and Herbert Wagner for helpful discussions.

References

XVIII.1 R. Lipowsky, J. Appl. Phys. **55**, 2485 (1984).

XVIII.2 R. Lipowsky, *The semi–infinite Potts model* (in german), PhD Thesis, University of Munich 1982 (unpublished); Z. Phys. B **45**, 229 (1982).

XVIII.3 R. Lipowsky, Phys. Rev. Lett. **49**, 1575 (1982) and Z. Phys. B **51**, 165 (1983).

XVIII.4 V. S. Sundaram, B. Farrell, R. S. Alben, and W. D. Robertson, Phys. Rev. Lett. **31**, 1136 (1973).

XVIII.5   R. Lipowsky and W. Speth, Phys. Rev. B **28**, 3983 (1983).

XVIII.6   E. G. McRae and R. A. Malic, Surf. Sci. **148**, 551 (1984).

XVIII.7   S. F. Alvarado, M. Campagna, A. Fattah, and W. Uelhoff, Z. Phys. B **66**, 103 (1987).

XVIII.8   H. Dosch, L. Mailaender, A. Lied, J. Peisl, F. Grey, R. L. Johnson, and S. Krummacher, Phys. Rev. Lett. **60**, 2382 (1988).

XVIII.9   See, e.g., A. R. Ubbelhode, *The Molten State of Matter*, (John Wiley, New York 1978).

XVIII.10   J. W. M. Frenken and J. F. van der Veen, Phys. Rev. Lett. **54**, 134 (1985).

XVIII.11   B. Pluis, A. W. Denier van der Gon, J. W. M. Frenken, and J. F. van der Veen, Phys. Rev. Lett **59**, 2678 (1987).

XVIII.12   K. C. Prince, U. Breuer, and H. P. Bonzel, Phys. Rev. Lett. **60**, 1146 (1988), and U. Breuer, H. P. Bonzel, K. C. Prince, and R. Lipowsky, Surf. Sci. (submitted).

XVIII.13   D. M. Zhu and J. G. Dash, Phys. Rev. Lett. **57**, 2959 (1986).

XVIII.14   J. Krim, J. P. Coulomb, and J. Bouzidi, Phys. Rev. Lett. **58**, 583 (1987).

XVIII.15   M. Bienfait, Europhys. Lett. **4**, 79 (1987).

XVIII.16   D. M. Zhu and J. G. Dash, Phys. Rev. Lett. **60**, 432 (1988).

XVIII.17   E. G. McRae and R. A. Malic, Phys. Rev. Lett. **58**, 1437 (1987).

XVIII.18   For simplicity, it is tacitly assumed here that $V(\ell)$ has a unique minimum. Beyond mean–field theory, $V_{DI}(\ell)$ must be replaced by a renormalized interaction which includes the effect of interfacial fluctuations. Quite generally, these fluctuations act to separate the (do) interface from the (vd) interface and, thus, to enhance the tendency towards complete surface–induced disorder (or surface melting).

XVIII.19   See, e.g., J. S. Rowlinson and B. Widom, *Molecular Theory of Capillarity* (Clarendon, Oxford 1982).

XVIII.20   D. M. Kroll and R. Lipowsky, Phys. Rev. B **28**, 6435 (1983).

XVIII.21   R. Lipowsky, Phys. Rev. Lett. **57**, 2876 (1986); Ferroelectrics **73**, 69 (1987).

XVIII.22   R. Lipowsky, *Critical behavior of interfaces: wetting, surface melting and related phenomena*, Habilitations–Thesis, University of Munich 1987, published as Jül–Spez–438 by KFA Jülich, Postfach 1913, D–5170 Jülich, West Germany.

XVIII.23   The length scales ξ_d and κ_d are usually microscopic. However, if the bulk transition is *weakly* first–order, the length scale ξ_d is set by the bulk correlation length and, thus, can be mesoscopic. In a fluid context, such a situation occurs, e.g., at the transition from a nematic to an isotropic liquid, see R. Lipowsky and D. Sullivan, Phys. Rev. Lett. **60**, 242 (1988).

XVIII.24 R. Lipowsky, D. M. Kroll, and R. K. P. Zia, Phys. Rev. B **27** 4499 (1983).

XVIII.25 D. M. Kroll and G. Gompper, Phys. Rev. B **36**, 7078 (1987).

XVIII.26 See, e.g., J. N. Israelachvili, *Intermolecular and Surface Forces* (Academic, Orlando 1985).

XVIII.27 J. L. Morán–López, F. Mejía–Lira, and K. H. Bennemann, Phys. Rev. Lett. **54**, 1936 (1985).

XVIII.28 R. Lipowsky, U. Breuer, K. C. Prince, and H. P. Bonzel, Phys. Rev. Lett. **62**, 913 (1989).

XVIII.29 In a more detailed model, one may decompose $\tilde{M}(\vec{Q}'' = \vec{0}, z)$ into several components which describe the decay of translational order *perpendicular* to the surface. Then the dominant part of $\tilde{M}(\vec{Q}'' = \vec{0}, z)$ can be an OP density rather than an NO density, which would lead to the critical behavior as given by (9).

XVIII.30 B. Pluis, T. N. Taylor, D. Frenkel, and J. F. van der Veen (to be published).

XVIII.31 R. Lipowsky, Phys. Rev. Lett. **52**, 1429 (1984); Z. Physik B **55**, 345 (1984).

XVIII.32 A. Trayanov and E. Tosatti, Phys. Rev. Lett. **59**, 2207 (1987).

XVIII.33 H. Löwen, T. Beier, and H. Wagner (to be published), and G. Gompper and D. M. Kroll (to be published) have derived this form from density functional theories for surface melting.

XVIII.34 For a recent example, see B. D. Swanson, H. Stragier, D. J. Tweat, and L. B. Sorensen, Phys. Rev. Lett. **62**, 909 (1989).

XVIII.35 H. Löwen, (to be published); H. Löwen and T. Beier (to be published).

XVIII.36 R. Lipowsky and M. E. Fisher, Phys. Rev. B **36**, 2126 (1987).

XVIII.37 K. Binder and D. P. Landau, Phys. Rev. B **37**, 1745 (1988).

XVIII.38 G. Gompper and D. M. Kroll, Phys. Rev. B **38**, 459 (1988).

XVIII.39 S. Dietrich and H. Wagner, Z. Phys. B **56**, 207 (1984).

XIX. Theory of Spin Waves in Magnetic Multilayers

J. Mathon

Department of Mathematics, City University, Northampton Square,
London EC1V 0HB, UK

A general method for calculating the local density of spin wave states in an arbitrary ferromagnetic multilayer is described. The method is very simple, computationally stable and so accurate that multilayers containing up to $10^3 - 10^4$ atomic planes can be easily handled. The results for the local DOS and the temperature dependence of the local magnetization in magnetic surfaces and superlattices are presented. The local magnetization obeys a Bloch $T^{3/2}$ law with a pre-factor that varies from one atomic plane to another and is determined by local exchange interactions. The temperature dependence of the local magnetization can, therefore, be used as a sensitive probe of local exchange interactions in any magnetic multilayer. First-principle formulation of the spin-wave problem for a magnetic overlayer above a non-magnetic metallic substrate is also given.

XIX.1. Introduction

Magnetic multilayers are now in the forefront of research in magnetism. They are very interesting not only because of their potential applications in perpendicular recording [XIX.1] but also because one can stabilize via epitaxy new physical systems that do not otherwise occur in nature [XIX.2].

There are three main factors that govern the magnetic properties of multilayers:
  (i) layer–by–layer magnetic moment in the ground state;
 (ii) layer–by–layer exchange interactions;
(iii) layer–by–layer anisotropy.
It is essential for design of novel magnetic multilayers that we fully understand and can control all these factors.

The ground state is now quite well investigated [XIX.3] and the main challenge facing both theory and experiment is, therefore, to determine the local exchange and anisotropy of magnetic multilayers.

XIX.2. Temperature Dependence of the Magnetization as a Probe of Local Exchange Interactions

One way to gain access to local exchange interactions is via spin waves. The general idea of the method is as follows. A model exchange Hamiltonian H

Springer Proceedings in Physics, Vol. 50   **Magnetic Properties of Low-Dimensional Systems II**
Editors: L.M. Falicov · F. Mejía-Lira · J.L. Morán-López © Springer-Verlag Berlin, Heidelberg 1990

for a layer structure is assumed. The local spin–wave Green function $G = (E - H)^{-1}$ in the mixed Bloch–Wannier representation $G_{ii}(E, q_{\parallel})$ is computed and the density of spin–wave states (DOS) in any atomic plane $i = 1, \ldots, N$ of the structure is obtained from $\mathrm{Im}\, G_{ii}$

$$N_i(E) = (1/\pi)\,\mathrm{Tr}\,\mathrm{Im}\,G_{ii}(E, q_{\parallel}), \qquad (XIX.1)$$

where the trace is over the wavevector $q_{\parallel}$ parallel to the structure surface. The local magnetization $M_i(T)$ is then given by

$$M_i(0) - M_i(T) = \int_0^{\infty} 2\mu_B \, N_i(E)[\exp(E/kT) - 1]^{-1}\, dE. \qquad (XIX.2)$$

Finally, the exchange parameters in the trial Hamiltonian are deduced from the best fit of Eq. (XIX.2) to the observed $M_i(T)$.

There are three crucial steps in this program. First it has to be demonstrated that variation of exchange over an interatomic distance $\simeq a$ leads to $M(T)$ that varies on the same microscopic scale. It is far from obvious that this is the case since $M(T)$ is determined by spin waves whose wavelengths are much longer than a. One also needs an accurate theory that can predict $M(T)$ in every atomic plane of an arbitrary multilayer. Finally, an experimental probe of $M(T)$ with a spatial resolution $\simeq a$ is required.

The two theoretical problems formulated above are addressed in Secs. XIX.2.1–XIX.2.3 and completely solved. The experimental probes capable of the required spatial resolution are discussed in the experimental contributions to this workshop (see also [XIX.3,XIX.4,XIX.5]).

XIX.2.1 Method of Adlayers

A high accuracy in the calculation of the local spin–wave density of states is essential. This is because the local magnetization is determined by spin waves whose energies are only of the order of 5–10% of the spin–wave band width.

The method of adlayers we have developed [XIX.6,XIX.7] can be viewed as the theoretical equivalent of an MBE machine. We start with a homogeneous semi–infinite ferromagnetic substrate with exchange J and spin S whose surface spin–wave Green function (surface DOS) is known from the classical spin–wave theory [XIX.6,XIX.7]. The next step is to "deposit" one–by–one all the atomic layers of the structure we wish to study. With every new adlayer the surface Green function is updated. At every stage of the deposition process we have, therefore, the exact surface Green function G^S expressed in terms of the known substrate Green function G^0. When the structure is thick enough for its magnetic properties to stabilize, we stop the deposition process and evaluate the local Green function in any atomic layer from G^S.

It is sufficient to describe the method for a simple cubic lattice and a nearest–neighbour exchange Hamiltonian since any layer structure can be reduced to this problem using the concept of principal layers [XIX.8].

The general recursion step from a layer n to the next layer $n+1$ is deduced from the Dyson equation $G^{n+1}=G^n+G^n W G^{n+1}$, where G^n is the old surface Green function, G^{n+1} the updated Green function and W is a 2×2 perturbation matrix due to deposition of the adlayer $n+1$. The recursion step takes the form

$$(G_{n+1\,n+1})^{-1} = \omega + W_{n+1\,n+1} - W^2_{n\,n+1} G_{n\,n} (1 - W_{n\,n} G_{n\,n})^{-1}, \qquad (XIX.3)$$

where ω is the diagonal matrix element of the substrate Hamiltonian and W depends on the local spin and exchange (see [XIX.7]). Once G^s is known, the surface DOS is determined from Eq. (XIX.1) where the BZ sum over $q_{\parallel}$ is performed using the Cunningham points [XIX.9]. The number of sampling points required in the calculation of the DOS is of the order of 2000 in the irreducible wedge of the BZ.

XIX.2.2 Magnetic Overlayer

Our hypothesis that the variation of local exchange on an atomic scale is reflected in the local $M(T)$ can be now tested for a ferromagnetic surface.

Classical spin–wave theory [XIX.10] predicts that the surface magnetization $M_S(T)$ obeys a Bloch law

$$M_S(T)/M_S(0) = 1 - kCT^{3/2} \qquad (XIX.4)$$

with $k = 2$ (C is the bulk prefactor).

Measurements of $M_S(T)$ for ferromagnetic metals using SPLEED [XIX.11], Mössbauer spectroscopy [XIX.12] and low–energy cascade electrons [XIX.13] confirm that M_S obeys a $T^{3/2}$ law but with a prefactor which is different for different surfaces and can be as large as $k = 5.4$ [XIX.13].

It is natural to assume that a faster decrease of $M_S(T)$ is due to a softening of surface exchange. One can clearly model such a softening by an overlayer with a suitable distribution of exchange deposited on a homogeneous ferromagnetic substrate.

Expanding Eq. (XIX.3) to the lowest order in E, it can be shown that the exact *initial* DOS in the surface of an overlayer consisting of N at. layers is given by [XIX.6,XIX.7]

$$N_S(E) = (S_N/S_{N-1})\ldots(S_1/S)2N_B(E) = (S_N/S)2N_B(E) + O(E^{3/2}), \qquad (XIX.5)$$

where $N_B(E) \propto E^{1/2}$ is the bulk DOS and S_N is the surface spin. The most remarkable feature of the surface DOS in Eq. (XIX.5) is that it is quite inde-

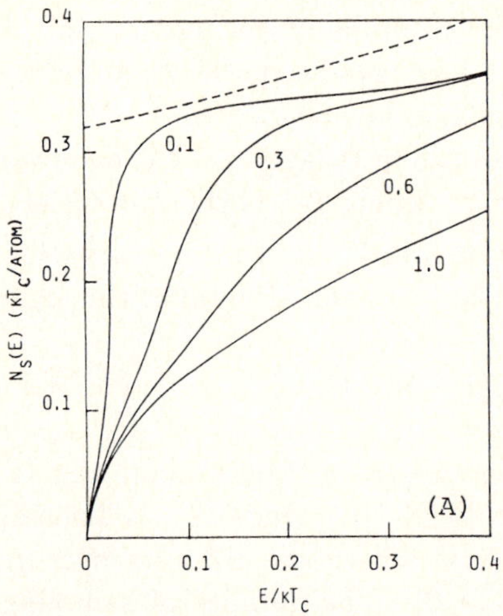 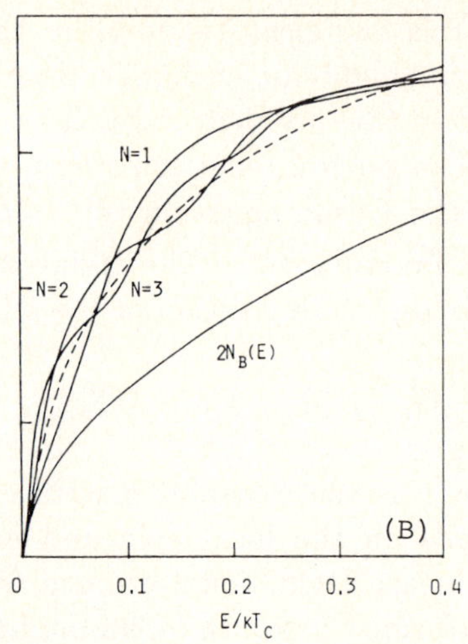

Figure XIX.1. Density of spin–wave states for a surface plane with surface to bulk exchange $J_\perp/J$=0.1, 0.3, 0.6, 1.0 (A) and for an over–layer of N planes with perpendicular exchange $J_\perp/J$=0.3 (B). The broken curve in (A) is for a 2D ferromagnet.

pendent of the exchange in the overlayer. It would appear from this exact result that no information about local exchange can be obtained from the Bloch law for the surface magnetization. Moreover, normalization to $M_s(0)$ in Eq. (XIX.4) means that the surface spin is cancelled and we recover from Eqs.(XIX.2) and (XIX.5) the classical result $k = 2$, which contradicts experiment.

Fortunately, quite a different picture emerges when the surface DOS is computed over a wider energy range. This is illustrated in Fig.XIX.1a for an overlayer with $N = 1$ (single at. layer) coupled to the substrate by a weaker exchange $J_\perp < J$. It can be seen that the classical result (XIX.5) breaks down at energies as low as $0.01kT_C$. The surface DOS at experimentally relevant energies is determined entirely by the surface exchange $J_\perp$.

Given that there is no good reason for $N_s(E)$ to be proportional to $E^{1/2}$ beyond the classical region (XIX.5), it remains to be explained why the observed $M_s(T)$ obeys a good $T^{3/2}$ law. Clue lies in the behaviour of the surface DOS for a thicker overlayer. The DOS for an overlayer of N at. layers ($N = 1, 2, 3, ..., 8$) with perpendicular exchange $J_\perp = 0.3J$ between the layers is shown in Fig. XIX.1b. Contrary to one's expectations, it follows from Fig. XIX.1b that an overlayer as thin as one at. layer already assumes the identity of an infinite overlayer with perpendicular exchange J (broken curve in Fig. XIX.1b). Since $M_s(T)$ for an infinite overlayer obeys naturally a Bloch law, it follows from Fig. XIX.1b that the same law must hold also for any overlayer with $N = 1, 2, 3, ..., 8$. This is demonstrated in Fig. XIX.2 which shows the calculated $M_s(T)$ for a range of $J_\perp$ together with the SPLEED results for FeNiB glass [XIX.11].

170

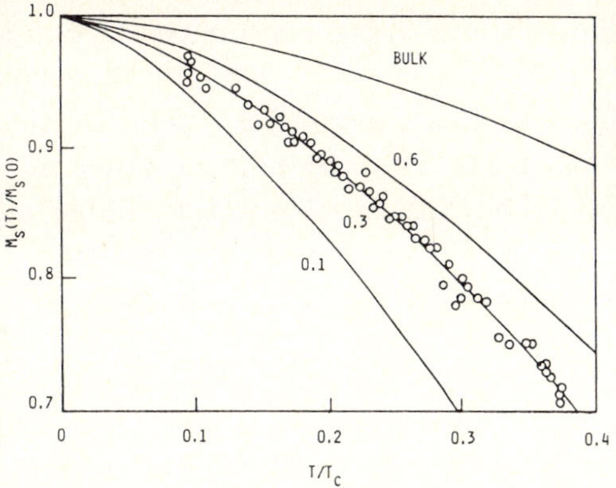

Figure XIX.2. Temperature dependence of the surface magnetization for surface to bulk exchange $J_\perp/J$=0.1, 0.3, 0.6. Circles are the SPLEED results for FeNiB glass [XIX.11].

Perfect fit to the experimental points is obtained for $J_\perp/J$=0.3, which means that the surface to bulk exchange in this sample is only 30% of the bulk J.

Recent measurements [XIX.14] with low–energy cascade electrons on the FeNi surface with a thin Ta interlayer confirm that the surface magnetization behaves precisely as predicted in Fig. XIX.2.

We have thus demonstrated that the temperature dependence of the local magnetization can be used as a sensitive probe of local exchange interactions in ferromagnetic surfaces and overlayers.

XIX.2.3 Magnetic Superlattice

The method of adlayers is so fast and accurate that we can model an infinite superlattice simply by depositing a sufficient number of complete superlattice periods on a substrate [XIX.15]. In practice, it is better to use a symmetric sandwich structure. A number Np of complete periods is first deposited. This is followed by an incomplete period containing $1 < n < N$ at. planes, where N is the superlattice period. We call such a structure the left (L) overlayer. The right (R) overlayer also consists of Np complete periods but the last incomplete period contains only $N - n$ at. planes. Finally the two overlayers are joined together. The local Green function in the plane next to the cleavage plane is computed from the surface Green functions $G^{L)}$ and $G^{(R)}$ for the two overlayers using the Dyson equation.

As an example, consider a superlattice with a weak exchange link $J_\perp$ =0.1J inserted after every third at. plane. The parallel exchange is J in all planes. This is a superlattice analog of the NiFe–Ta–NiFe sandwich with a weak exchange link which was studied in reference [XIX.14].

The choice of a substrate is important. Using the adlayer method, it can be shown analytically [XIX.15] that the exact initial DOS for the above superlattice is that of an anisotropic ferromagnet with perpendicular exchange $J_\perp^{-1}$ =(1/3)[1/J+1/J+1/0.1J] and parallel exchange J. This is the best choice for a substrate since it will enforce on our finite structure the exact initial DOS for the infinite superlattice.

There are two inequivalent at. planes in this superlattice. One has the weak link on one side only (1) and the other has the bulk links on either side (2). The corresponding DOSs are shown in Fig. XIX.3a. The total number of complete periods used in this calculation is 20–50, which is quite sufficient to model an infinite superlattice. The broken curve in Fig. XIX.3a is the exact initial DOS.

As for overlayers, the exact initial DOS, which is common to all at. planes, is valid only at energies so low that they are of no practical interest. On the other hand, the local DOS reflects closely local exchange in the superlattice unit cell. In particular, the steep jump in the DOS obtained for planes of type 1 should be clearly observable through a sudden much more rapid decrease of $M(T)$ in these planes which sets in at $T \simeq 0.25 T_C$.

Since the inner atomic planes are inaccessible to most experimental probes, we have also calculated the local DOS for a semi–infinite superlattice with different planes from its unit cell exposed at the surface. There are three inequivalent surface planes and their DOSs are shown in Fig. XIX.3b. All three DOSs are now different but the characteristic jump in the DOS obtained for the infinite superlattice remains and should be measurable by surface probes of $M_s(T)$.

We conclude that the temperature dependence of the local magnetization contains detailed information not only about the local exchange in a superlattice but also about the composition of its unit cell (via the location of jumps in DOS).

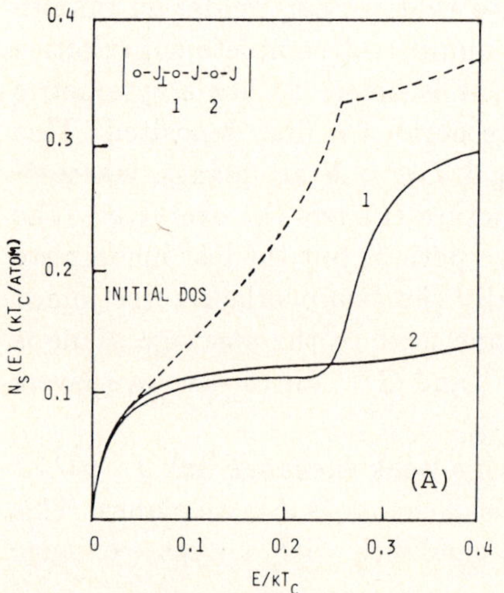

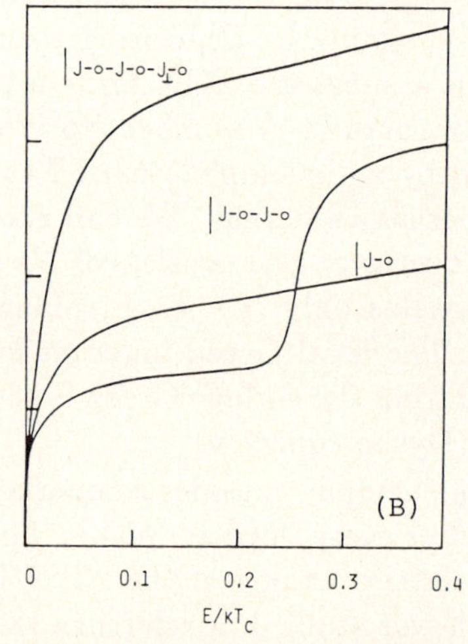

Figure XIX.3. Local densities of spin–wave states for all inequivalent at. planes in an infinite superlattice with a weak exchange link $J_\perp /J = 0.1$ inserted after every third at. layer (A) and for a semi–infinite superlattice with different planes from its unit cell exposed at the surface (B).

It is also clear that the local $M(T)$ in a superlattice can be manipulated by a suitable choice of the materials in its unit cell. This could be important for magneto–optic recording which relies on a local softening of $M(T)$ induced by heating with a laser pulse.

XIX.3. First–Principles Calculation of Spin Waves in an Overlayer

The ultimate theoretical goal is to solve from first principles the spin–wave problem for an arbitrary multilayer. We have implemented such a calculation for a transition–metal overlayer modelled by a simple non–degenerate band with a strong intra–atomic repulsion U (one–band Hubbard model).

The calculation starts with a non–magnetic ($U{=}0$) metallic substrate whose surface electron Green function G^0 is determined by the transfer–matrix method [XIX.16]. An overlayer of N at. planes is then deposited and all the matrix elements of its spin–dependent one–electron Green function G^σ_{ij} are generated from the substrate G^0 using the adlayer method. The interaction U is switched on and the ferromagnetic ground state of the overlayer is calculated in the Hartree–Fock (HF) approximation. All the matrix elements of the HF transverse susceptibility $\chi^{HF}_{ij}(\omega, q_\parallel)$ in the overlayer are then computed from the HF one-electron Green function G^σ [XIX.3].

The ground is now prepared to attack the spin–wave problem. We recall that spin waves are poles of the full dynamic susceptibility χ^{-+} which is given in the random phase approximation (RPA) by [XIX.3] $\chi^{-+} = (I - U\chi^{HF})^{-1}\chi^{HF}$. The energies ω of all spin–wave modes in the overlayer can be thus determined from the secular equation

$$\mathrm{Det}\,|I - U\chi^{HF}(\omega, q_\parallel)| = 0, \qquad (XIX.6)$$

where I is the $N \times N$ unit matrix.

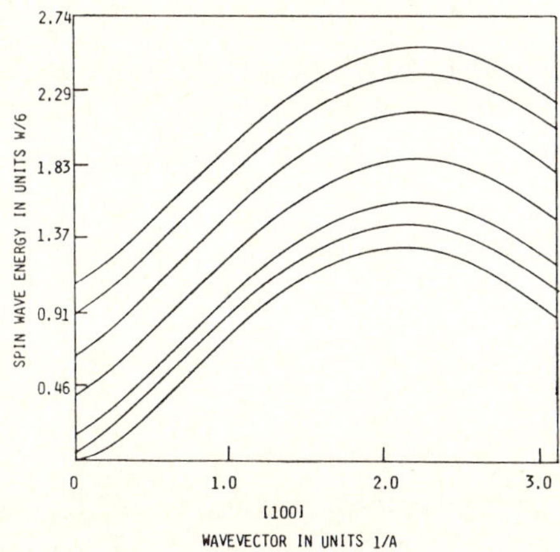

Figure XIX.4. Energies of all spin–wave modes in an overlayer of seven at. planes on a (100) non–magnetic metallic substrate.

As an illustration, all the spin–wave modes in an overlayer of 7 at. planes
with 0.12 holes in its spin majority band and with an average exchange splitting
$\Delta \simeq 0.5w$ are shown in Fig. XIX.4 (w is the band width).

Since the method of adlayers used in the above calculation is applicable to
an overlayer of arbitrary composition, the spin–wave problem for the overlayer
with a single–orbital band structure is completely solved. Generalisation to
a fully realistic multi–orbital band structure via the principal layer concept
is straightforward and work is now in progress on spin waves in Ni and Fe
overlayers.

References

XIX.1 S. Iwasaki, IEEE Trans. Magn. MAG–**20**, 657 (1984).

XIX.2 A. S. Arrott, B. Heinrich, S. T. Purcell, J. F. Cochran, and K. B.
Urquhart, J. Appl. Phys. **61**, 3721 (1987); S. D. Bader and E. R. Moog,
J. Appl. Phys. **61**, 3729 (1987).

XIX.3 J. Mathon, Rep. Prog. Phys. 51, 1 (1988).

XIX.4 L.M. Falicov and J.L. Morán-López (eds.), *Magnetic Properties of Low-
Dimensional Systems*, Springer Verlag, Berlin (1986).

XIX.5 R. Feder (ed.), *Polarized Electrons in Surface Physics*, World Scientific,
Singapore (1985).

XIX.6 J. Mathon and S. B. Ahmad, Phys. Rev. B **37**, 660 (1988).

XIX.7 J. Mathon, Physica B **149**, 31 (1988).

XIX.8 D. H. Lee and J. D. Joannopoulos, Phys. Rev. B **23**, 4988 (1981).

XIX.9 S. L. Cunningham, Phys. Rev. B **10**, 4988 (1974).

XIX.10 G. T. Rado, Bull. Am. Phys. Soc. **2**, 127 (1957); D. L. Mills and A. A.
Maradudin, J. Phys. Chem. Solids **28**, 1855 (1967).

XIX.11 D. T. Pierce, R. J. Celotta, J. Unguris, and H. C. Siegmann, Phys.
Rev. B **26**, 2566 (1982).

XIX.12 J. C. Walker, R. Droste, G. Stern, and J. Tyson, J. Appl. Phys. **55**,
2500 (1984); J. Korecki and U. Gradmann, Phys. Rev. Lett. **55**, 2491
(1985).

XIX.13 D. Mauri, D. Scholl, H. C. Siegmann, and E. Kay, Phys. Rev. Lett. **61**,
758 (1988).

XIX.14 H. C. Siegmann, D. Mauri, D. Scholl, and E. Kay, J. Phys. (Paris) **49**,
C8-9 (1988).

XIX.15 J. Mathon, J. Phys.: Condensed Matter **1**, 2585 (1989).

XIX.16 M. P. López Sancho, J. M. López, and J. Rubio, J. Phys. F **15**, 851
(1985).

XX. Spin Fluctuation Theory of Surfaces, Sandwiches and Superlattices

H. Hasegawa

Department of Liberal Arts, Tohoku Gakuin University,
Izumi-ku, Sendai 981-31, Japan

A review is given of our recent calculations on surfaces, sandwiches and superlattices by using the local spin–fluctuation theory which was proposed and developed by the author. We discuss magnetic and electronic properties of 1) free surfaces of Fe(001), Ni(001), and Cr(001); 2) Ni thin films sandwiched by Cu, and 3) Co/Cr superlattices including exchange anisotropy.

XX.1. Introduction

Research on magnetism of transition–metal surfaces and interfaces is one of the most active fields in solid state physics. In recent years considerable progress has been made in our theoretical understanding of surface–related problems. The sophisticated local–spin density functional (LSDF) method provides a good description for the ground state properties of surfaces, interfaces and superlattices of transition metals (for a review, see Ref. XX.1). When we pay attention to their finite–temperature properties, they have been so far investigated mainly by localized (Heisenberg or Ising) models. Since most systems are composed of transition metals, it is desirable to study their magnetic properties based on the itinerant–electron or band model.

In the last decade there has been significant development in the theory of finite–temperature bulk band magnetism. It was a long–standing problem how to reconcile the itinerant and localized characters observed experimentally in transition metals and alloys. The former manifests itself in non–integral magnetic moments, the Fermi surface, etc. and the latter in the Curie–Weiss susceptibility and large specific–heat peaks near the Curie or Néel temperature. Paying attention to the importance of the effect of local magnetic moments, which is neglected in the conventional Hartree–Fock or Stoner theory, Hasegawa [XX.2] proposed a local spin–fluctuation theory for finite–temperature band magnetism based on the Hubbard model; an equivalent theory was independently proposed by Hubbard [XX.3]. In this approach the Stratonovich–Hubbard functional integral method is employed within the static approximation to include the effect of local spin fluctuations, which is very important in discussing magnetism at finite temperatures. It is a mean–field theory in which the system under consideration is regarded as a collection of local magnetic moments. This approach has proved particularly useful in understanding magnetic properties at finite temperatures

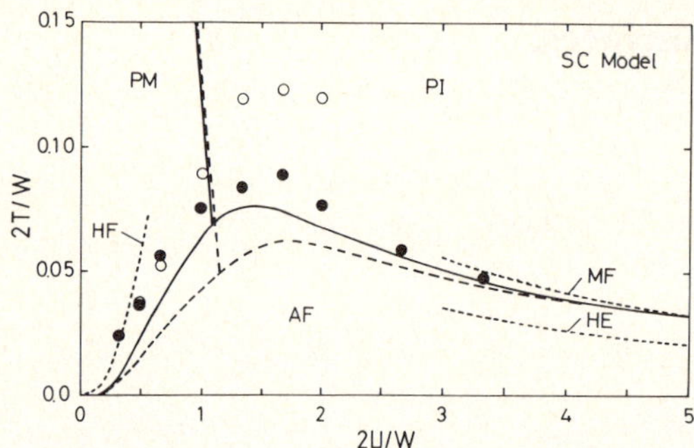

Figure XX.1. The U–T phase diagram of the simple–cubic half–filled Hubbard model showing the antiferromagnetic (AF) state, paramagnetic metal (PM) and paramagnetic insulator (PI), calculated by the HH theory (solid curves [XX.2]), the GVA theory (dashed curves [XX.5]) and MC simulations (circles, [XX.6,XX.7]). Dotted curves denote T_N in the Hartree–Fock (HF) theory and molecular–field (MF) and high–temperature expansion (HE) approximations for the Heisenberg model [XX.4].

[XX.4]. We show in Fig. XX.1 the Néel temperature, T_N, as a function of the interaction strength, U, of the simple–cubic Hubbard model, which was calculated by various methods. We note that the result of the Hasegawa–Hubbard (HH) local spin–fluctuation theory is in quantitatively good agreement with the Monte Carlo simulations [XX.6,XX.7]; a related discussion is given in Sec. 6. This justifies to a large extent the HH theory, which has been applied to various problems of transition metals [XX.8].

It is indispensable to take into account the effect of local spin fluctuations in discussing surface magnetism of transition metals. Quite recently the HH theory has been successfully extended and applied to surface–related problems [XX.9–XX.12]. It is the purpose of this paper to discuss our recent calculations in conjunction with relevant experimental results. In the following section (Sec. 2) we briefly mention the generalization of the HH local spin–fluctuation theory to surface–related problems. In Sec. 3 we discuss the free surfaces of Fe(001), Ni(001) and Cr(001). Section 4 deals with thin Ni films sandwiched by Cu. The exchange anisotropy, discussed in Sec. 5, is based on our calculation of Co/Cr superlattices. The final section is devoted to conclusions and supplementary discussion.

XX.2. Local Spin–Fluctuation Theory

We employ a semi–infinite *fcc* or *bcc* lattice with the (001) surface (or interface). The layers parallel to (001) are labelled by the index n, which is 1 for the surface. The direction of spin quantization is taken in the z axis although it is independent of the surface (or interface) direction since we neglect crystalline anisotropy. The model Hamiltonian is expressed by [XX.9]

$$H = H_\circ + H_\iota, \qquad (XX.1)$$

with

$$H_\circ = \sum_\sigma \sum_j \sum_m E_j\, a^\dagger_{jm\sigma} a_{jm\sigma} + \sum_\sigma \sum_{j,j'} \sum_{m,m'} t^{m\,m'}_{jj'}\, a^\dagger_{jm\sigma} a_{j'm'\sigma'}, \qquad (XX.2)$$

$$H_\iota = (1/4) \sum_j \left(U_j\, N_j^2 - J_j\, M_j^2 \right). \qquad (XX.3)$$

The first term denotes the tight–binding d–band Hamiltonian and the second term the interaction term. In eqs. (XX.1)–(XX.3) $a_{jm\sigma}$ is an annihilation operator of the σ–electron of orbital m at site j, and $N_j\,(M_j)$ denotes the total charge (magnetic moment) operator. We adopt parametrized hopping integrals, $t_{jj'}$, of canonical band theory to get realistic, degenerate d–bands. The surface was introduced by cutting the hopping integrals across a cleavage plane. All atomic potentials, E_j, were set to their bulk values except for E_1 on the surface which was chosen to simulate the surface moments consistent with the LSDF calculations for slabs. The exchange and Coulomb interactions, J_j and U, on all layers were assumed to be equal to their bulk values and reproduce results in the ground–state band calculations. The explicit parameter values adopted in our calculations are given in Refs. [XX.9–11].

The model Hamiltonian of Eq. [XX.1] was used by many authors to investigate surface magnetism within the Hartree–Fock approximation. We take into account the effect of local spin fluctuations, which is completely neglected in the Hartree–Fock approximation, by means of the HH theory [XX.2,XX.3]. We adopt the Stratonovich–Hubbard functional–integral method within the static approximation. The partition function of the interacting electron system is expressed as the partition functions of non–interacting systems in random charge (ν_j) and exchange (ς_j) fields with Gaussian weight. The former field is treated by the saddle–point approximation and the latter is included in the alloy–analogy approximation by means of the coherent potential approximation (CPA). Self–consistent equations for the mean and root–mean–square (RMS) exchange fields, $\langle \varsigma \rangle_n$ and $\langle \varsigma^2 \rangle_n$, and the average charge field, $\langle \nu \rangle_n$, on the n–th layer are given by [XX.9a]

$$\langle \varsigma \rangle_n = \int d\varsigma_j\, \varsigma_j\, C_n(\varsigma_j), \qquad (XX.4)$$

$$\langle \varsigma^2 \rangle_n = \int d\varsigma_j\, \varsigma_j^2\, C_n(\varsigma_j), \qquad (XX.5)$$

$$\langle \nu \rangle_n = -i \int d\epsilon\, f(\epsilon)(-1/\pi)\, Im \sum_\sigma F_{n\sigma}(\epsilon). \qquad (XX.6)$$

The relevant equations for the *bulk* case are given by Eqs. (XX.4)–(XX.6), suppressing the index n. The major complication in surface related problems is that the exchange and charge fields depend on the layer n. The n–th layer is regarded as a multi–component "random alloy" with concentration distribution $C_n(\varsigma_j)$, for the potential of $(1/2)J_n\,\varsigma_j$ and is characterized by the n–dependent coherent potentials, $\Sigma_{n\sigma}(\epsilon)$. The local Green's function of a σ–spin electron on the layer n, $F_{n\sigma}(\epsilon)$, is calculated by the transfer matrix method [XX.9b]. Note that $C_n(\varsigma_j)$ and $F_{n\sigma}(\epsilon)$ depend on $\langle\varsigma\rangle_n$, $\langle\varsigma^2\rangle_n$, and $\langle\nu\rangle_n$ through $\Sigma_{n\sigma}(\epsilon)$. When the self–consistent equations are solved, we can obtain physical quantities such as the mean and RMS values of magnetic moments, $\langle M\rangle_n$ and $\langle M^2\rangle_n^{1/2}$, and the number of electrons, $\langle N\rangle_n$, on layer n [XX.9a].

In the conventional Hartree–Fock theory, $\langle M\rangle_n$ and $\langle N\rangle_n$ are determined self–consistently. The present theory determines not only $\langle M\rangle_n$ and $\langle N\rangle_n$ but also $\langle M^2\rangle_n$ (spin fluctuations) in a self–consistent way. The effect of local spin fluctuations on all layers as well as in the bulk are treated on the same footing. The theory can interpolate between the weak– and strong–coupling limits. In the weak–coupling limit (or at $T = 0$ K) the present theory reduces to the molecular–field approximation for the Heisenberg model. Just as finite–temperature band theory has been useful in understanding various problems of bulk magnetism [XX.8], our theory is meaningful also for surface–related problems, as is shown in the following sections.

XX.3 Free Surfaces

XX.3.1 Fe(001)

The study of magnetism of Fe surfaces has a relatively long history, beginning with an observation of magnetic "dead" layers on Fe surfaces. This early result is now believed to be caused by contaminated Fe surfaces. A recent LSDF calculation [XX.13] showed that surface moments are enhanced by about 30% compared with the bulk moment of 2.2 μ_B. This is consistent with Mössbauer experiments [XX.14].

The temperature dependence of layer magnetizations calculated by the present theory is shown in Figure XX.2 [XX.9c]. The surface moment, $\langle M\rangle_l$ which is enhanced compared with the bulk at $T = 0$ K, decreases much faster than that in the bulk upon heating. This is consistent with angle–resolved photoemission (ARPE) [XX.15,XX.16] and spin–polarized low–energy–electron diffraction SPLEED [XX.17]. A smaller surface coordination number leads to a narrower surface band and to enhanced surface moments at $T = 0$ K. At elevated temperatures, however, it yields a faster decrease in the surface magnetization because of enhanced surface spin fluctuations. The surface perturbation is essentially confined to the first layer at $T = 0$ K. With rising temperature, it penetrates deeply and affects interior layers.

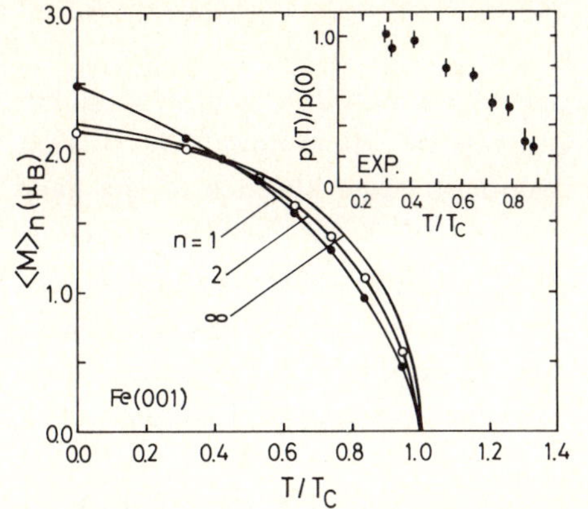

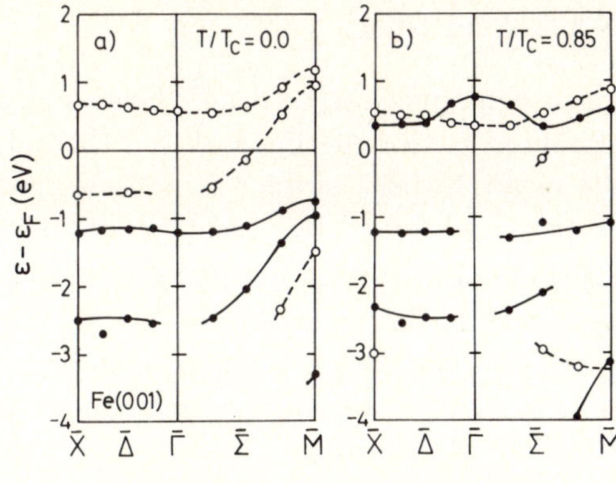

Figure XX.2. The calculated temperature dependence of the layer magnetization, $\langle M \rangle_n$, of Fe(001) [XX.9c]. The inset shows the normalized spin polarization observed by ARPE [XX.15].

Figure XX.3. Calculated surface energy bands of Fe(001) at (a) $T/T_C = 0$ and (b) 0.85 [XX.9c]. Filled and open circles denote peak positions in the up–spin and down–spin spectral densities, respectively. Curves are drawn only to guide the eye.

One of the advantages of the present theory is that we can discuss the temperature dependence of the electronic structure which can be observed, for example, by ARPE experiments. We show in Fig. XX.3 the temperature dependence of the surface band structure of Fe(100) [XX.9c]. It shows the contours of peak positions in the calculated surface spectral densities along the high–symmetry lines in the surface Brillouin zone. It includes not only the surface states (or resonances) but also non–surface (bulk) states. At $T = 0$ K the surface spectral densities are exchange split, with almost uniform exchange splitting. When the temperature is raised, the exchange splitting of $\bar{\Gamma}$–point spectral densities decreases and it produces a broad single peak at and above T_C. On the other hand, the peak positions in the surface states at the middle of the $\bar{\Gamma}$–$\bar{\Sigma}$–$\bar{M}$ and $\bar{\Gamma}$–$\bar{\Delta}$–$\bar{X}$ lines are almost temperature independent. Thus the surface spectral densities of Fe(001) show a complicated behavior which depends on the surface wave vector.

The situation is quite similar in bulk spectral densities of Fe. It has been shown that the peak position at Γ'_{25} is only weakly temperature dependent while the exchange splitting of states near the H point vanishes above T_C [XX.9c]. This is in qualitative agreement with the observed ARPE spectra [XX.15,XX.16], but it is in strong contrast with the behavior of Ni, as discussed below.

XX.3.2 Ni (001)

Since a magnetically "dead" layer was reported on Ni(001), this system has attracted both experimental and theoretical attention. It is now believed that the clean Ni(001) surface is not magnetically dead. The absence of the dead layer was observed by recent experiments such as SPLEED [XX.18], electron–capture spectroscopy (ECS) [XX.19] and ARPE [XX.20]. This was supported by recent LSDF calculations [XX.21] which showed the surface moment to be slightly (about 20%) enhanced compared with the bulk moment of $0.6\mu_B$. This enhancement is attributed partly to a narrower surface band and partly to surface dehybridization.

The temperature dependence of layer magnetizations of Ni(001) calculated by our theory is very similar to that of Fe(001) [XX.9d], as Fig. XX.4 shows. The surface moment, enhanced at $T = 0$ K, decreases much faster than that in the bulk and shows an almost linear dependence near the Curie temperature. These results are confirmed by ECS [XX.22], ARPE [XX.23] and SPLEED [XX.24].

It is found that the exchange–split peaks in surface as well as bulk spectral densities for all wavevectors merge into single peaks above T_C [XX.9d]. This result is supported by ARPE data [XX.22].

XX.3.3 Cr(001)

The magnetic properties of Cr surfaces are quite different from that of bulk Cr, which is in a SDW state with a maximum moment of 0.59 μ_B at $T = 0$ K. The possibility of a peculiar magnetic state in Cr surface was pointed out by several theoretical calculations [XX.25,XX.26]. Recent LSDF calculations [XX.27] showed that Cr(001) is ordered ferromagnetically with an enhanced surface moment of about 2.5 μ_B .

Fig. XX.5 shows the temperature dependence of layer magnetizations of Cr(001) calculated by our theory [XX.9b]. We assumed bulk Cr to be in the

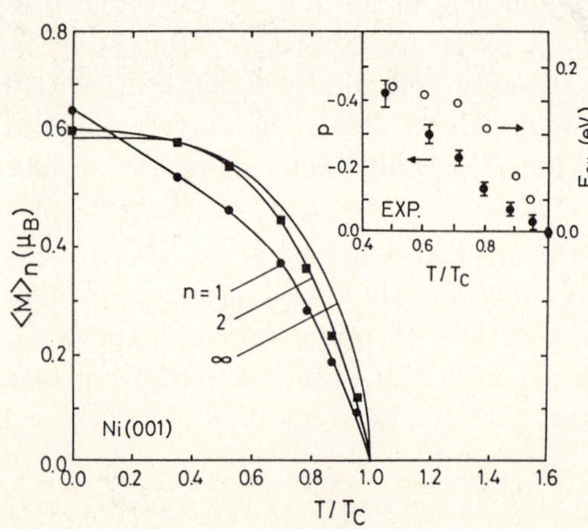

Figure XX.4. The calculated temperature dependence of the layer magnetization, $\langle M \rangle_n$ of Ni(001) [XX.9c]. The inset shows the normalized spin polarization observed by ECS (•, [XX.19]) and ARPE (○, [XX.20]).

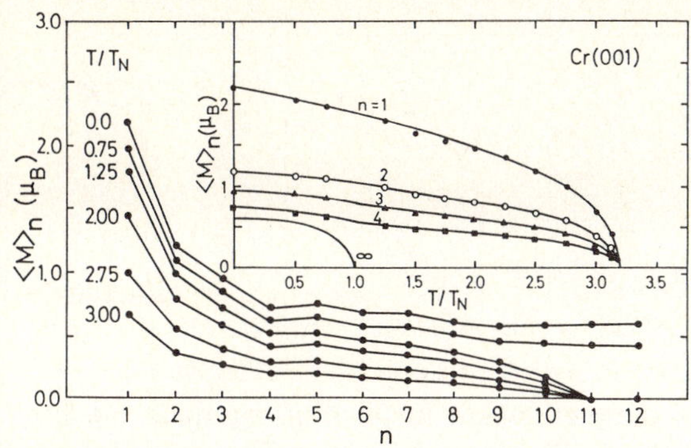

Figure XX.5. Calculated magnetization profiles in Cr(001), $\langle M \rangle_n$, the sign of magnetization changes layer to layer [XX.9b]. The inset shows $\langle M \rangle_n$ against temperature.

antiferromagnetic ground states with the sublattice moment of 0.6 μ_B . At $T = 0$ K the ferromagnetic order is realized at the surface with an enhanced moment, as in previous calculations [XX.25,XX.26]. The effect of a significant surface perturbation penetrates deeply, at least up to the 6– or 7–th layer. It was shown that surface magnetic order persists up to the surface Curie temperature of $T_{CS} = 3.2T_N$.

There is currently a consensus on the theoretical picture for Cr surfaces. It is, however, not the case in the observed results. The ARPE experiment on Cr(001) [XX.26] reported the persistence of the surface magnetic order up to 780 K $(T_{CS} \sim 2.5T_N)$, which is consistent with our result of $T_{CS} = 3.2T_N$. On the contrary, the recent SPLEED [XX.29] showed the absence of magnetic order on Cr(001). In order to reconcile these contradicting results, Blügel *et al.* [XX.30] proposed a model in which adjacent terraces of ferromagnetic Cr(001) planes are assumed to couple antiferromagnetically. Further theoretical as well as experimental studies are required to clarify this confusing issue.

XX.4. Ni Thin Films

There has been considerable experimental and theoretical work on the magnetism of thin Ni films. We show in Figs. XX.6 and XX.7 some recent experimental data of the ground–state magnetic moment, $M(0)$, and the Curie temperature, T_C , as functions of the Ni film thickness, D. They show that both $M(0)$ and T_C decrease with decreasing D.

We adopted in our calculation a Cu/Ni/Cu sandwich where a D–layer Ni film is on a semi–infinite Cu(001) and is covered by a five–layer Cu overlayer [XX.10].

The calculated ground–state moment, $\langle M(0) \rangle$, shown in Fig. XX.6, becomes smaller for thinner Ni films. This is in semi–quantitative agreement

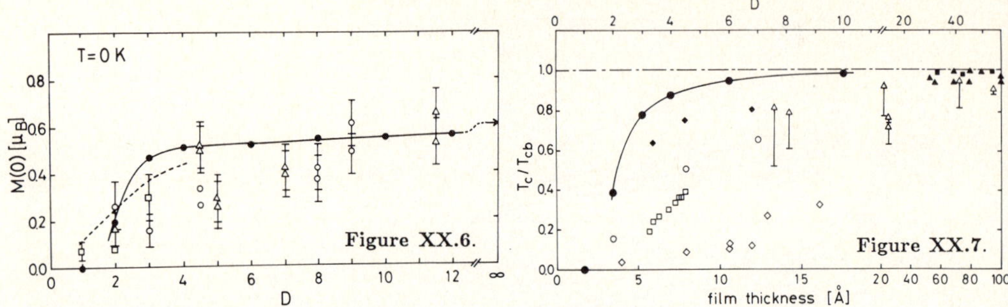

Figure XX.6. The ground–state magnetization in Ni films against the film thickness, D, the arrow denoting the bulk value [XX.10]. The dotted curve expresses the LSDF calculation for Ni–Cu superlattices [XX.31]. Experimental data are for Cu/Ni/Cu(001) sandwiches with different experimental conditions [XX.32].

Figure XX.7. The normalized Curie temperature, T_C/T_{Cb}, of Ni films as a function of film thickness, D [XX.10]. T_{Cb} denoting the bulk Curie temperature. Experimental data are for Cu/Ni/Cu(001) (○ [XX.32]), Au/Ni/Au(111) (□ [XX.33]), compositionally modulated Ni–Cu superlattices (◇,◆, [XX.34,XX.35]), and earlier ones on thicker Ni films (■, △, ▲, [XX.36]).

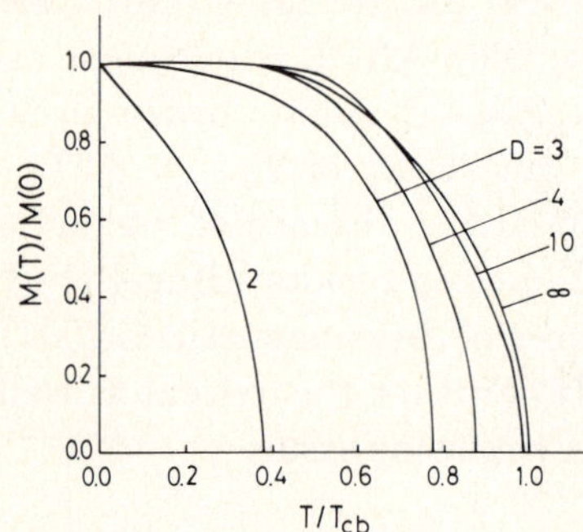

Figure XX.8. The normalized magnetization, M(T)/M(0), as a function of the reduced temperature, T/T_{Cb} for various Ni film thickness D.

with the LSDF calculation for Ni/Cu superlattices [XX.31] and with experimental data [XX.32]. Fig. XX.8 shows the calculated temperature dependence of the normalized magnetization for various film thickness. The variation of $M(T)/M(0)$ against the temperature becomes significant as D is reduced. This kind of behavior is widely observed in various systems such as Ni/Cr [XX.37] and Fe/V superlattices [XX.38]. The calculated D–dependences of $M(0)$ and T_C for Ni thin films are shown in Figs. XX.6 and XX.7, respectively. The calculated trends, showing smaller $M(0)$ and T_C for thinner Ni films, are qualitatively consistent with experimental results [XX.32,XX.39,XX.40]. Existing data are, however, significantly scattered because of experimental difficulties for preparing ultra–thin films.

XX.5. Co/Cr Superlattices (Exchange Anisotropy)

Since the discovery of exchange anisotropy in Co–CoO in 1957, ferromagnetic–antiferromagnetic (F/AF) interfaces have been the source of both scientific and technological interest. One of the manifestiations of the exchange anisotropy is a shift in the B-H curve, which is referred to as the effective exchange field, H_{ex}. There has been a controversy on the magnitude of H_{ex} and the origin of its peculiar temperature dependence, which shows an almost linear decrease up to the Curie temperature.

In order to get a microscopic understanding of F/AF interfaces, we consider a Co/Cr superlattice as a prototype of F/AF systems. We applied our theory to four–layer (1Co–3Cr) and eight–layer (3Co–5Cr) superlattices with bcc (001) interfaces [XX.11]. The calculated temperature dependence of layer magnetizations, $\langle M \rangle_n$, in 3Co–5Cr superlattice is shown in Fig. XX.9. At $T = 0$ K Co magnetic moments are ferromagnetically ordered while Cr moments are in the antiferromagnetic states. This is consistent with the LSDF calculation for the Co/Cr Brillouin curve. On the other hand, the $\langle M \rangle_n$ curves of Cr layers show a very peculiar temperature dependence. This was shown as due to a fairly weak exchange interaction across the Co/Cr interfaces, J_{CoCr}. We should note that if J_{CoCr} vanishes, the Co and Cr layers have distinct transition temperatures of T_C (Co) and T_n (Cr). When a small J_{CoCr} exists, the Cr moments near the Co/Cr interface can be polarized, even at T_N (Cr) $< T < T_C$ (Co) because of the exchange field arising from Co layers through the weak exchange coupling.

The temperature dependence of H_{ex} and the exchange anisotropy constant, I, are calculated by using the expression given by [XX.11]

$$H_{ex} = 2I/[\sum_{n \in F} \langle M \rangle_n],\qquad (XX.7)$$

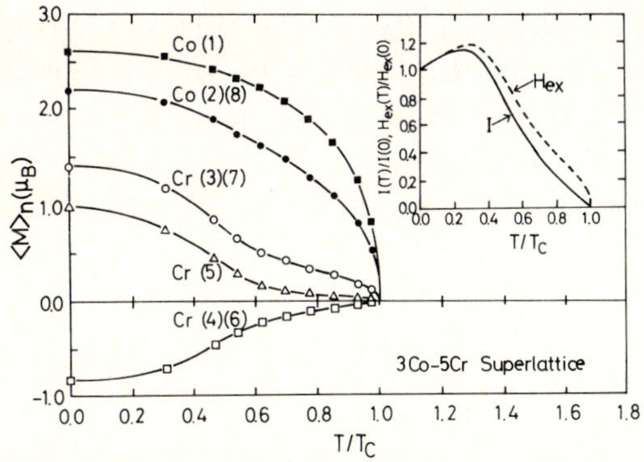

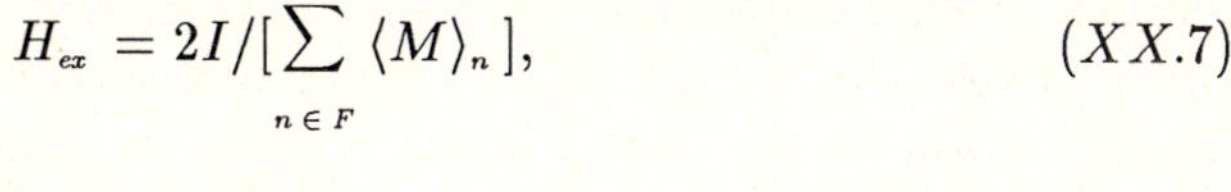

Figure XX.9. The calculated temperature dependence of the layer magnetization, $\langle M \rangle_n$, in the eight–layer (3Co–5Cr) superlattice. The inset shows the theoretical temperature dependence of the effective exchange field, H_{ex}, and the exchange anisotropy constant, I [XX.11].

with

$$I = J_{CoCr} \langle M \rangle_{Co\,l} \langle M \rangle_{Cr\,l}, \qquad (XX.8)$$

where n denotes the F layer and $\langle M \rangle_{Co\,l}$ ($\langle M \rangle_{Cr\,l}$) is the moment of the interface Co (Cr) layer. The calculated results are plotted in the inset of Fig. XX.9. Both I and H_{ex} show an unusual temperature dependence, which deviates significantly from the Brillouin function and which shows nearly linear decrease above $T/T_C \sim 0.5$. This is mainly due to the peculiar temperature dependence of interfacial Cr moment. Such a peculiar temperature dependence is widely observed in many F/AF interface systems. Our calculation shows a peak in I (and H_{ex}) at $T/T_C \sim 0.3$, which arise from the combined effects of an increase in $J_{Co\,Cr}$ and a decrease in $\langle M \rangle_{Cr\,l}$ when the temperature is raised (see Fig. 11 in Ref. XX.11b). Such a peak was observed in some F/AF systems, for example NiFe/Mn [XX.42], but not in others such as FeMn/FeNi [XX.43]. The absence of the peak would be realized if both $J_{Co\,Cr}$ and $\langle M \rangle_{Cr\,l}$ decrease monotonically upon heating.

XX.6. Conclusions and Discussions

The magnetic properties of Cr surface are unique compared with those of Fe and Ni surfaces. Cr surfaces have a ground–state moment which is 4~5 times larger than that in the bulk. This is understood based on the fact that the Fermi level is located between the up–spin and down–spin surface states, and these surface states work to enhance the Cr surface moment. In the case of Fe and Ni, the surface states are not effective in enhancing surface moments, because both up–spin and down–spin surface states are located below the Fermi level. Then, the ground–state surface moments in Fe and Ni are only slightly enhanced due to the narrowing of the surface band. The local moments on the Cr surface above T_N are much larger (1.6~2.0 μ_B) than in the bulk and this is the origin of the enhanced surface Curie temperature of $T_{C\,s} \sim 2.5 T_N$. This is easily understood based on a mean–field type equation given by

$$T_C = \max(T_{Cb}, T_{Cl}), \qquad (XX.9)$$

with

$$T_{Cb} = z_b J_b \langle M^2 \rangle_b /3, \qquad T_{Cl} = z_S J_S \langle M^2 \rangle_S /3 + J_b \langle M^2 \rangle_b /3, \qquad (XX.10)$$

where z is the coordination number, J the exchange interaction and the subscript b and S stand for the bulk and surface, respectively. The enhanced local moment: $\langle M^2 \rangle_S \gg \langle M^2 \rangle_b$, leads to $T_{CS} > T_{Cb}$ for the Cr surface. In the case of Fe and Ni, we get $T_{CS} = T_{Cb}$ because the reduced surface coordination number, $z_S < z_b$, overweights the slightly enhanced surface moments. Recently, a LSDF calculation predicted giant moments in thin Cr films on Cu and Ag substrates [XX.44]. A study of their stability at finite temperatures should be interesting.

It has been shown that the temperature dependence of the surface spectral densities of Fe and Cr depend on the surface wave vector, whereas that of Ni is almost the same throughout surface Brillouin zone. We note that this difference arises from the large difference in the surface moments. Local moments on Fe and Cr surfaces are about 2.0–2.5 μ_B , which are much larger than that on the Ni surface of $\sim 0.6\mu_B$. The calculated temperature dependence of spectral densities were in qualitative or semi–quantitative agreement with relevant ARPE results [XX.15,XX.16,XX.23.XX.28], although there remains a controversy on the extent of the short–range magnetic order (SRMO) and the role of the surface in the interpretation of ARPE data (see discussion in Ref. XX.1).

One of the disadvantages of HH theory [XX.2,XX.3], based on which our calculations have been performed, is that it cannot include the effect of electron correlations. Quite recently, the present author proposed a new theory by using a Gutzwiller–type variational approach (GVA) [XX.5]. In this approach the Kotliar–Ruckenstein functional integral method [XX.45] is combined with an alloy–analogy approximation [XX.2,XX.3]. The theory reduces, for vanishing temperature, to Gutzwiller's approximation [XX.46] while in the high–temperature limit it reduces to the HH theory. The GVA theory also describes sensible crossovers between the weak– and strong–coupling limits. We show by the dashed curve in Fig. XX.1, the Néel temperature of the simple–cubic model calculated by the GVA theory [XX.5]. It is lower than T_N in the HH theory because of the effect of electron correlations. The temperature dependence of sublattice magnetization of the simple–cubic model is shown in Fig. XX.10, where solid and dashed curves denote results in the GVA and HH theories, re-

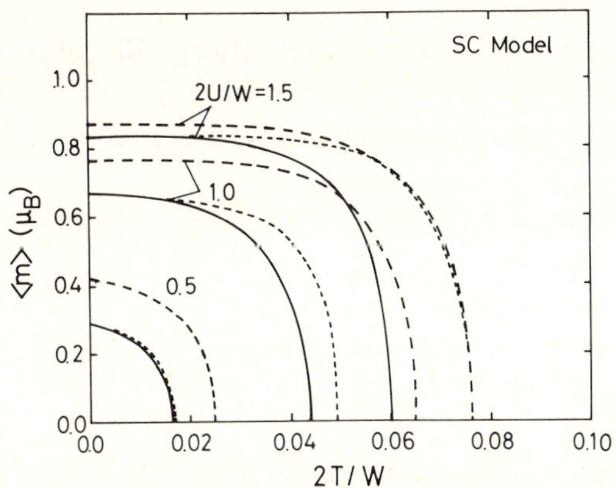

Figure XX.10. The temperature dependence of sublattice magnetizations of the simple–cubic model for various interactions, calculated in the GVA (solid curves [XX.5]), HH theory (dashed curves [XX.2,XX.3]), and HH theory with the renormalized interaction (dotted curves) (see text).

spectively. For a comparison, we show the dotted curve which gives the results obtained using the HH theory but with the renormalized interaction which was chosen to reproduce the ground–state sublattice magnetization obtained in the GVA theory. This suggests that the results calculated with the use of the HH theory have at least semi–quantitative validity. By using the newly developed theory [XX.5], we may be able to discuss magnetic properties of surface–related problems in more detail.

References

XX.1 J. Mathon, Progr. Theor. Phys. **51**, 1(1988).

XX.2 H. Hasegawa, J. Phys. Soc. Jpn. **46**, 1504 (1979); **49** 178 (1980).

XX.3 J. Hubbard, Phys. Rev. B **19**, 2626 (1979).

XX.4 Y. Kakehashi and H. Hasegawa, Phys. Rev. B **36**, 4066; **37**, 7777 (1988).

XX.5 H. Hasegawa, J. Phys.: Cond. Mat. (in press).

XX.6 J. E. Hirsch, Phys. Rev. B **35**, 1851 (1986).

XX.7 R. T. Scalettar, D. J. Scalapino, R. L. Sugar, and D. Toussaint, Phys. Rev. B **39**, 4711 (1989).

XX.8 Related references can be found in Ref. XX.4.

XX.9 H. Hasegawa, J. Phys. F **16**, 347 (1986);**16**, 1555 (1986); **17**, 165 (1987); **17**, 679 (1987).

XX.10 H. Hasegawa, Surf. Sci. **182**, 591 (1987).

XX.11 H. Hasegawa and F. Herman, Physica B **149**, 175 (1988); Phys. Rev. B **38**, 4863 (1988).

XX.12 H. Hasegawa and F. Herman, J. Phys. (Paris) **49**, 1677 (1988).

XX.13 S. Ohnishi, A. J. Freeman, and M. Weinert, Phys. Rev. B **28**, 6741 (1983).

XX.14 J. Tyson, A. H. Owens, and J. C. Walker, J. Appl. Phys. **52**, 2487 (1981).

XX.15 E. Kisker, K. Schroeder, W. Gudat, and M. Campagna, Phys. Rev. B **31**, 329 (1985).

XX.16 J. Kirschner, M. Globel, V. Dose, and H. Scheidt, Phys. Rev. Lett. **53**, 612 (1984).

XX.17 J. Kirschner, Phys. Rev. B **30**, 415 (1984).

XX.18 R. Feder, F. Alvarado, E. Tamura, and E. Kisker, Surf. Sci. **127**, 83 (1983).

XX.19 C. Rau and S. Eihner, Phys. Rev. Lett. **47**, 939 (1981).

XX.20 W. Eberhardt and E. W. Plummer, Phys. Rev. B **30**, 3113 (1984).

XX.22 C. Rau and H. Keffer, J. Magn. Magn. Mat. **54–57**, 767 (1986).

XX.23 R. Raue and H. Hopster, Z. Phys. B **54**, 121 (1984).

XX.24 S. Alvarado, M. Campagna, and H. Hopster, Phys. Rev. Lett. **48**, 51 (1982).

XX.25 Y. Teraora and J. Kanamori, *Transition Metals* (Inst. Phys. Conf. Ser. 39, 1977) p. 588.

XX.26 D. R. Grempel, Phys. Rev. B **24**. 3928 (1981).

XX.27 C. L. Fu and A. J. Freeman, Phys. Rev. B **33**, 1755(1986).

XX.28 L. E. Klebanoff, S. W. Robey, G. Liu, and D. A. Shirley, Phys. Rev. B **30**, 1048 (1984).

XX.29 S. F. Alvarado and C. Carbone, Physica B **149**, 43 (1988).

XX.30 S. Blügel, D. Pescia, and P. H. Dederichs, Phys. Rev. B **39**, 1392 (1989).

XX.31 A. J. Freeman, J. Xu, and T. Jarlborg, J. Magn. Magn. Mat. **31–34**, 909 (1983).

XX.32 L. R. Sill, M. B. Brodsky, S. Bowen, and H. C. Hamaker, J. Appl. Phys. **57**, 3663 (1985).

XX.33 P. Beauvillain, C. Chappert, J. P. Renard, C. Maliere, and D. Renard, preprint.

XX.34 W. S. Zou, H. K. Wong, J. R. Owens–Bradley, and W. P. Halperin, Physica B **108**, 953 (1981).

XX.35 J. Q. Zheng, J. B. Ketterson, C. M. Falco, and I. K. Schuller, J. Appl. Phys. **53**, 3150 (1982).

XX.36 See Ref. 10.

XX.37 M. B. Stearns and C. H. Lee, J. Appl. Phys. (in press).

XX.38 H. K. Wong, H. Q. Yang, J. E. Hilliard, and J. B. Ketterson, J. Appl. Phys. **57**, 3660 (1985).

XX.39 W. S. Zhou, H. K. Wong, J. R. Owens–Bradley, and W. P. Halperin, Physica B **108**, 953 (1981).

XX.40 F. Meier, D. Pescia, and T. Scrieber, Phys. Rev. Lett. **48**, 645 (1982).

XX.41 F. Herman, Ph. Lambin, and O. Jepsen, Phys. Rev. B **31**, 4394 (1985).

XX.42 O. Masenet, R. Montmory, and L. Néel, J. Appl. Phys. **52**, IEEE Trans. Magn. MAG–1, 63 (1965).

XX.43 C. Tsang, N. Heiman, and K. Lee, J. Appl. Phys. **52**, 2471 (1981); A. P. Malezemoff, Phys. Rev. B **35**, 3679 (1987).

XX.44 C. L. Fu, A. J. Freeman, and T. Oguchi, Phys. Rev. Lett. **54**, 2700 (1985); R. Richter, J. G. Gay and J. R. Smith, Phys. Rev. Lett. **54**, 2704 (1985).

XX.45 G. Kotliar and A. E. Ruckenstein, Phys. Rev. Lett. **57**, 1362 (1986).

XX.46 M. C. Gutzwiller, Phys. Rev. **137**, A1726 (1965).

XXI. Surface Magnetic Properties of a Transverse Ising Model with a Dilute Surface

E.F. Sarmento [1] *and T. Kaneyoshi* [2]

[1]Universidade Federal de Alagoas, Departamento de Fisica,
 57000 Maceio, Brasil
[2]University of Nagoya, Department of Physics, 464 Nagoya, Japan

Surface magnetic properties of a semi–infinite spin 1/2 Ising ferromagnet with a dilute surface in an applied transverse field are investigated by use of an effective–field theory with correlations. A number of characteristic behaviors appear in the temperature and transverse–fields dependences of the surface magnetization. The results suggest that research in surface magnetization in applied transverse fields at very low temperature may provide a new method for classifying experimentally some problems in surface magnetism.

XXI.1. Introduction

Surface physics and surface magnetism are rapidly growing fields (see Binder [XXI.1] and Diehl [XXI.2] for recent reviews). The standard theoretical example consists of spins which on the surface interact with each other via an exchange interaction J_s different from that in the bulk, J. The model exhibits different types of phase transitions associated with the surface; if the ratio $\Delta_s = J_s / J - 1$ is greater than a critical value Δ_c, the system may order at the surface before it orders in the bulk. The system exhibits two successive transitions, namely surface and bulk transitions, as the temperature is lowered. If the ratio is less than Δ_c, the system orders at the bulk transition temperature.

The purpose of this work is to investigate temperature (or transverse field) dependences of surface magnetization in a semi–infinite transverse Ising model with a dilute surface within the same framework as that of Sarmento *et al.* [XXI.3]. A number of characteristic behaviors appear in the temperature and transverse–field dependences of surface magnetization, depending on whether Δ_s is larger (or smaller) than a critical value Δ_c.

XXI.2. Formulation

We consider a model system with a dilute surface which is described by the following Hamiltonian:

$$H = -\frac{1}{2} \sum_{ij} J_{ij}\, s_i^z s_j^z\, \xi_i\, \xi_j - \Omega \sum_i s_i^x\, \xi_i \qquad (XXI.1)$$

where s_i^x and s_i^z are components of a spin–1/2 operator at site i, Ω represents

Springer Proceedings in Physics, Vol. 50   **Magnetic Properties of Low-Dimensional Systems II**
Editors: L.M. Falicov · F. Mejía-Lira · J.L. Morán-López © Springer-Verlag Berlin, Heidelberg 1990

the transverse field and the first summation is carried out only over nearest-neighboring pairs of spins. J_{ij} takes either the value J_S if both spins lie on the surface, or the bulk value J otherwise; ξ_i is a random variable which at the surface takes value 1 or 0, depending on whether or not a magnetic atom occupies a given surface site; in the bulk $\xi = 1$.

By following the procedure described in Ref. [XXI.3], the average longitudinal surface magnetization m_S^z can be expressed by

$$m_S^z = 2\langle\langle S_{i=s}^z \rangle\rangle_T = \{p[\cosh(DJ_S) + m_S^z \sinh(DJ_S)] + (1-p)\}^4$$

$$\times [\cosh(DJ) + m^z \sinh(DJ)]f(x)|_{x=0} \qquad (XXI.2)$$

where $D = \partial/\partial x$ is a differential operator and $\langle...\rangle_r$ denotes the random configurational average. The function $f(x)$ is defined by

$$f(x) = \frac{x}{[4x + x^2]^{\frac{1}{2}}} \tanh(\frac{\beta}{4}[4\Omega^2 + x^2]^{1/2}) \qquad (XXI.3)$$

where $\beta = 1/k_B T$. In addition, $p = \langle \xi_i \rangle_T$ is the concentration of magnetic atoms at the surface. For the magnetization of the first layer we have

$$m_1^z = 2\langle\langle s_{i=1}^z \rangle\rangle_T = \{p[\cosh(DJ) + m_S^z \sinh(DJ)] + (1-p)\}$$

$$\times [\cosh(DJ) + m_1^z \sinh(DJ)]^4 [\cosh(DJ) + m_2^z \sinh(DJ)]f(x)|_{x=0} \ . \qquad (XXI.4)$$

In general, the magnetization m_n^z of the n–th layer is given by

$$m_n^z = [\cosh(DJ) + m_n^z \sinh(DJ)]^4 [\cosh(DJ) + m_{n-1}^z \sinh(DJ)]$$

$$\times [\cosh(DJ) + m_{n+1}^z \sinh(DJ)]f(x)|_{x=0} \qquad (XXI.5)$$

where m_{n-1}^z and m_{n+1}^z are the magnetizations in the $(n-1)$–th and $(n+1)$–th layers, respectively.

We were unable to solve the above coupled equations analytically. Even if a numerical method is used, the sequence must be truncated after a given number of layers. The simplest approximation is to assume that the magnetization remains unaltered after the second layer, namely $m_2^z = m_3^z = ... = m_n^z = m_B^z$, where m_B^z is the bulk longitudinal magnetization. The transverse components m_S^x, m_1^x, and m_n^x are given by equations (XXI.2), (XXI.4), and (XXI.5) by changing $f(x) \to g(x)$, where $g(x)$ is given by

$$g(x) = \frac{2\Omega}{[4\Omega^2 + x^2]^{\frac{1}{2}}} \tanh(\frac{\beta}{4}[4\Omega^2 + x^2]^{\frac{1}{2}}) \ . \qquad (XXI.6)$$

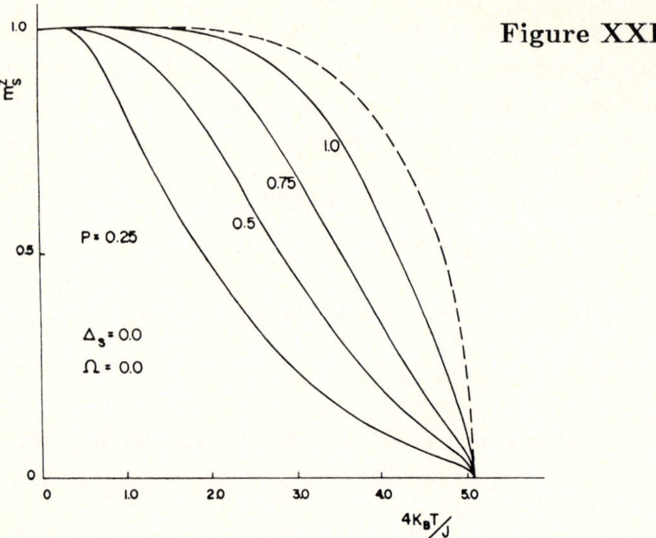

XXI.3. Dilute Surface

In this section, we study the effects of transverse field on the surface magnetization of the system with a dilute surface. In Fig. XXI.1 several curves of the temperature dependence of m_s^z for a system with $\Delta_s = 0.0$ are plotted. The dashed line represents the bulk magnetization curve m_B^z. Decreasing the concentration p of surface magnetic atoms, the magnetization shows a large decrease from m_s^z for $p = 1.0$ as well as from the bulk value m_B^z. The curve with $p = 0.25$ has a downward curvature in a wide region below $T = T_c^b$, since the surface is too dilute and the surface spins are essentially paramagnetic due to thermal fluctuations.

In Fig. XXI.2, curves of the temperature dependence of m_s^z for a system with $\Delta_s = 1.0$ are depicted. With the decrease of p the critical value of Δ_c

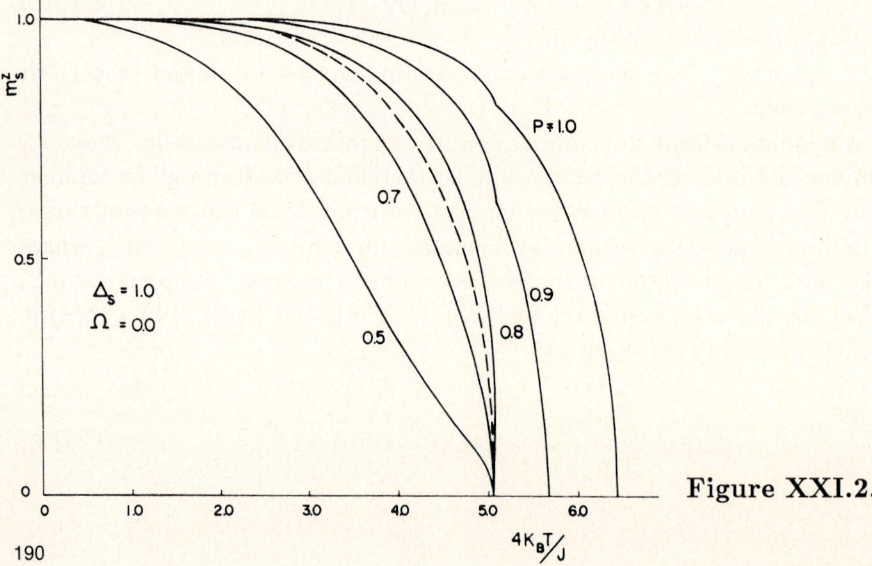

Figure XXI.2.

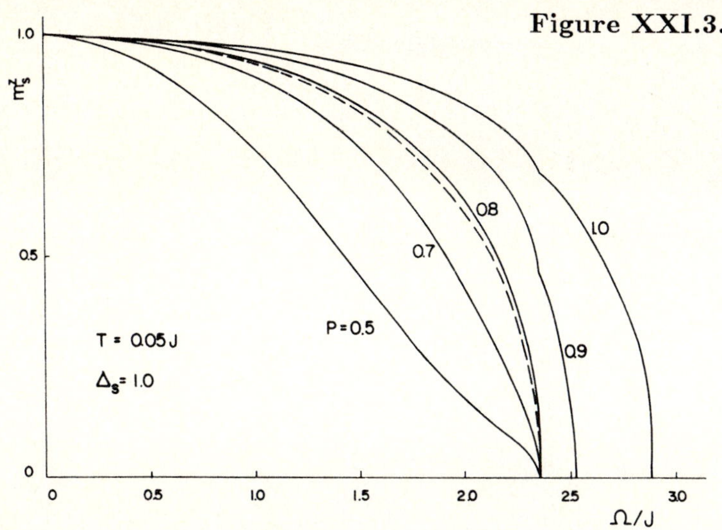

increases rapidly, and for p below $p = 0.7$ surface magnetic ordering can be achieved only at $T = T_C^b$, since the value of Δ_s becomes smaller than the critical value Δ_C. The dashed curve in the figure represents the bulk magnetization curve m_b^z. It can be seen that the magnetization curve m_s^z exhibits a small dip at $T = T_C^b$, which indicates that the first derivative of m_s^z at $T = T_C^b$ may have a discontinuity when surface magnetic ordering becomes possible above T_C^b. This discontinuity has also been obtained by renormalization–group theory [XXI.4]. The cluster–variation method [XXI.5] claims that the surface magnetization depends strongly on the boundary conditions imposed in surface calculations. Experimentally, results for Gd show no discontinuity at $T = T_C^b$ for either the surface magnetization and its first derivative [XXI.6].

In Fig. XXI.3, changes of m_s^z with Ω are shown for $\Delta_s = 1.0$, at very low temperatures ($T = 0.05J$) for various values of p. In the figure, solid and dashed lines represent the surface and bulk magnetization curves, respectively. Comparing Fig. XXI.2 with Fig. XXI.3, we find a similarity between the T–plot and the Ω–plot for surface magnetization. In Fig. XXI.3, the curve with $p = 0.9$ exhibits a small dip at the critical value $\Omega = \Omega_C$, since the critical value Δ_C is then smaller than $\Delta_s = 1.0$. Thus, determination of surface magnetization in an applied transverse field fixing T at very low temperature may be a powerful method for clarifying a variety of problems in surface magnetism experimentally.

The changes of m_s^z with Ω for the system with $\Delta_s = 0.0$, at $T = 0.05J$ and for various values of p are shown in Fig. XXI.4. By decreasing p the m^z curve changes from the linear behavior to an upward curvature, and the discontinuity of its first derivative at $\Omega = \Omega_C$ for $p = 1.0$ gradually disappears.

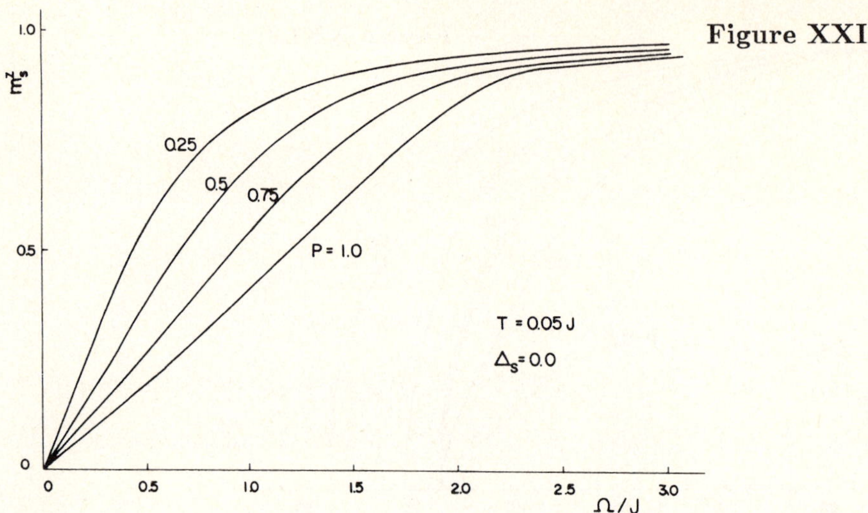

Figure XXI.4.

References

XXI.1 K. Binder, *Phase Transitions and Critical Phenomena*, Vol. 8, ed. by D. Domb and J. L. Lebowitz (Academic, New York, 1983).

XXI.2 H. W. Diehl, *Field–Theoretic Approach to Critical Behaviour at Surfaces*, Vol. 10, ed. by C. Domb and J. L. Lebowitz (Academic, New York, 1986).

XXI.3 E. F. Sarmento, I. Tamura, L. E. M. C. de Oliveira, and T. Kaneyoshi, J. Phys. C **17**, 3195 (1984).

XXI.4 C. Tsallis and A. Chame, J. Physique, in press.

XXI.5 J. L. Morán–López and J. M. Sánchez, preprint.

XXI.6 C. Rau and M. Robert, Phys. Rev. Lett. **58**, 2714 (1988).

Part 5

Clusters and One-Dimensional Magnetic Systems

XXII. Magnetic Interactions in One-Dimensional Systems

R.A. Barrio and Chumin Wang

Instituto de Investigaciones en Materiales, Universidad Nacional
Autónoma de México, Apartado Postal 70-360, 04510 México, D.F.

Self–consistent calculations for electrons in large linear chains which present quasicrystalline order of various kinds were performed. Peculiar properties of the magnetic behavior of these systems are discussed.

XXII.1. Introduction

Quasicrystals are rather fascinating systems in which the translational symmetry is absent. Nevertheless, they retain certain properties usually associated with crystals. Real quasiperiodic solids have been found in one, two and three dimensions and have produced an enormous amount of theoretical work in recent years.

The possibilities of technological applications of these unique systems are unknown so far, since, as we show below, the characteristics of the excitations in quasicrystals are very similar to the ones in systems which present two incommensurate periodicities, which have been known for a long time. Properties usually associated with the translational symmetry of crystals, for instance, Van–Hove singularities [XXII.1], k–conservation selection rules, Brillouin zone folding, etc. are also present in Fibonacci superlattices [XXII.2]. The interesting point is that we may be on the threshold of a new way of looking at ordered solids, and an opportunity to generalize some of the basic concepts of Solid State theory. This is so because quasiperiodic systems can be derived from strictly periodic structures in higher dimensions [XXII.3]. This is an essential issue to be addressed in this contribution.

From a very general point of view, quasicrystals might be expected to behave very similarly to normal crystals in experiments which probe quantities that are averaged over large regions in real space, except when long range correlations are important. An example of such an experiment is polarized Raman scattering, which probes the totally coherent vibrational modes in the solid. In a former contribution [XXII.4] we presented a study of the Raman response of Fibonacci superlattices made with crystalline layers of binary compounds. The calculated spectra compare extremely well with experiment and, although the effects of quasiperiodicity in the distribution of the Raman active modes are evident, the spectrum could be interpreted by using the picture of acoustic phonon folding as in normal periodic superlattices. Therefore one has to look harder

Springer Proceedings in Physics, Vol. 50   **Magnetic Properties of Low-Dimensional Systems II**
Editors: L.M. Falicov · F. Mejía-Lira · J.L. Morán-López © Springer-Verlag Berlin, Heidelberg 1990

for physical properties which are peculiar to quasicrystals in order to find new interesting phenomena.

The purpose of the present contribution is two fold: On one hand we would like to stress the similarity of the spectra of excitations in quasiperiodic chains and in incommensurate periodic chains, and on the other hand to present peculiar magnetic properties of the one–dimensional quasicrystals which arise from many–body interactions.

XXII.2. Electrons in Quasicrystals

When dealing with quasicrystals, one shall use quasiperiodic functions to describe the interactions between excitations. The definition of a quasiperiodic function is a straightforward generalization of periodicity [XXII.5].

A function of a real variable $f(x)$ is said to be quasiperiodic if, and only if, for any $\epsilon > 0$ there is a number $\tau(\epsilon, f)$ such that

$$\mid f(x + \tau) - f(x) \mid \leq \epsilon,$$

for all x. As an example consider the function

$$f(x) = \sin(2\pi x) + \sin(2\pi \sigma x), \qquad (XXII.1)$$

where $\sigma = (1+\sqrt{5})/2$ is the *golden section*. This function is obviously non-periodic, since there is not a real number such that $f(x + T) = f(x)$. However, equation (XXII.1) fulfills the definition of quasiperiodicity.

The properties of quasiperiodic functions have been fully analyzed [XXII.6], in particular, they can be obtained through a generalized Fourier transform. In general, one defines

$$F(x_1, x_2, ...) = \sum A_l \, exp(i \sum_m r_{lm} \, a_m \, x_m), \qquad (XXII.2)$$

where r_{lm} are rational numbers and a_m are linearly independent real numbers in the sense that if $\sum_m r_m a_m = 0$, then all r's are nil. The multiperiodic function F would, in general, have incommensurate periods, if the a's are irrational, therefore one could define any quasiperiodic functions as the diagonal part of a multiperiodic function in a space of more dimensions, for instance

$$f(x) = diag[F(x_1, x_2, ...)] = F(x, x, ...). \qquad (XXII.3)$$

The definition (XXII.3) is interesting, since it suggests that one could regard a quasiperiodic function of D variables as the projection of periodic points in a hyperspace on N dimensions into a hypersurface of D dimensions [XXII.3]. The Fibonacci sequence is one of the simplest examples of this procedure, because one needs only one irrational (σ) to project a periodic function in two dimensions onto one, as in equation (XXII.1). From equation (XXII.3) a number of properties of quasiperiodic functions can be derived [XXII.7].

Another very important conclusion that could be inferred from the definitions above is that one can expect that the properties of excitations in quasicrystals must retain the periodic nature of the parent function in a higher dimension. Let us restrict ourselves to the Fibonacci chain from now on in the understanding that many of the properties derived here for it should be applicable to other quasiperiodic systems.

There are various methods for contructing a Fibonacci chain. For instance, one could define a transformation rule that could be applied repeatedly to a seed [XXII.8], or one could add up chains of lower generations to form a longer chain [XXII.9]. Here the latter method was found convenient, since it provides a clear notion of the generation number n, the number of sites contained in each generation $m(n)$ and the key features for the renormalization procedure followed to describe the excitations.

The only two ingedients needed are: two sorts of objects and a rule to accommodate them in a sequence of sites. If one has already built chains up to generation $n-1$, the rule to build up the next generation is $C_n = C_{n-1} + C_{n-2}$, in which the order of addition, left or right, must be respected and the two first generations $n=0$ and $n=1$ contain a single object of each sort.

Let us consider an s–type tight–binding Hamiltonian, with nearest neighbour interactions only:

$$H = \sum_i E_i \mid i\rangle\langle i \mid + \sum_i t_{ij} (\mid i\rangle\langle i+1 \mid + \mid i+1\rangle\langle i \mid). \qquad (XXII.4)$$

Then, the two sorts of sites in the sequence could have two different site–energies E_A and E_B (site problem), or alternatively, one could define a bond problem by assigning two values to the hopping integrals t_l and t_s that give a long bond (l) and a short bond (s) respectively.

The equations of motion for the electron states $\mid i\rangle$ have been solved by using a renormalization procedure fully described elsewhere [XXII.10]. The key point of the procedure is to renormalize the central point of a chain of generation $n=2$ and built the next generation with these renormalized objects. The end result is that one finds easy recursion relations for the renormalized site energies and hopping integrals, and one is able to consider very large chains, since the number of sites $m(n)$ grows exponentially at each iteration.

In Figure XXII.1a the local density of states (DOS) at the central site of a chain of generation $n=50$ is plotted for the values $t_l=-1.0$ and $t_s=-1.5$. It can be noticed that the structure of the DOS is very similar to the ones reported for shorter chains [XXII.11] and for chains with periodic boundary conditions [XXII.12], except that the structure is much more accurately defined. A small imaginary part (of the order of 10^{-4} times the bandwidth) has been added to the

196

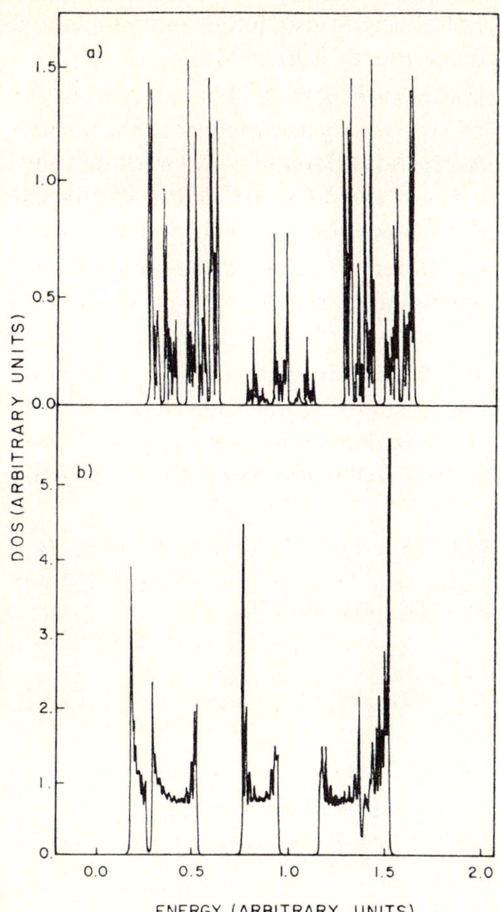

Figure XXII.1. a) Local density of states at the central atom of a Fibonacci chain of generation 50. b) Local density of states of the spin–up Hubbard sub–band taken from the central atom of a chain of 100 atoms with coordination 2 and Bethe lattices as saturators at the extremes. The number of electrons per atom is 0.8148.

energy. The fractal structure of the spectrum is evident since the distribution of gaps and states is self–similar [XXII.10].

We would like to stress here the similarity of the spectrum shown in Figure XXII.1a for the electrons in a Fibonacci chain to the one obtained in a normal periodic chain when a further periodicity, incommensurate with the lattice, is introduced. One simple way of illustrating this is to consider the behaviour of the chain when electron–electron correlations are modelled and the number of electrons in the band is not exactly one half, but some irrational number. An easy way to deal with the problem is to consider an extended Hubbard model and to use Hartree–Fock to linearize it. In this manner the problem is completely mapped into an effective tight–binding Hamiltonian, which may be solved self–consistently [XXII.13]. If one plots the array of bands and gaps, due to the Peierls instabilities, as a function of filling one obtains a self–similar pattern [XXII.14]. In Figure XXII.1b electronic spectrum of a chain with a band filled up to 0.4074 electrons is shown. Notice that the states follow a complicated distribution of bands and gaps, similar to the one in Figure 1(a).

It is clear that the similarity of the spectra in Figure XXII.1 arises from the fact that both reflect the superposition of two nearly incommensurate periodicities, however, one should expect more essential differences. Take for instance the site problem. In the Fibonacci chain there are only two values of the site energy which are placed in a non–periodic fashion in the chain, whilst in the self–consistent solution there are a large number of different site energies due to the electron correlations. This fact must have an influence in the spatial extension of the wave functions.

The localization problem becomes important, since in the linear chain all states become localized under infinitesimal amounts of disorder [XXII.15]. On the other hand, in the case of the Fibonacci diagonal problem it has been shown [XXII.16] that there are no exponentially localized states, but all of them are critical [XXII.17].

A way of investigating the localization properties of electron states in randomly disordered systems has been to calculate the Lyapunov exponents, defined as [XXII.18]

$$\gamma(E) = \lim_{m \to \infty} (\frac{1}{m} \ln \|M(m; E)\|), \qquad (XXII.5)$$

where $\| \ \|$ represents the norm of a matrix and $M(m; E)$ is a product of 2×2 transfer matrices $M_i(E)$, that are determined by the Fibonacci sequence. These transfer matrices are defined between the coefficients c_i forming the Wannier functions from a given site basis

$$\begin{pmatrix} c_{m+1} \\ c_m \end{pmatrix} = \prod_{i=1}^{m} M_i(E) \begin{pmatrix} c_1 \\ c_0 \end{pmatrix} = M(n, E) \begin{pmatrix} c_1 \\ c_0 \end{pmatrix}. \qquad (XXII.6)$$

In the bond problem M_i could be of three different forms [XXII.19], because there are only three different pairs of bonds in the sequence. It is clear that, when there is periodicity, the product of matrices in (XXII.6) is identically 1 for the period and $\gamma=0$ for the extended states characterized by the eigenvalue k of the group of translations. On the other hand, by looking at equation (XXII.6), it is seen that if the product tends to infinity, the difference between c_m and c_1 is also infinite, and if the former is finite, the latter must be zero, meaning that there is no contribution to the wave function from very distant sites, and therefore the state is localized. In particular, if the product of matrices grows exponentially, then the state becomes exponentially localized.

One could calculate a good approximation to equation (XXII.5) by using a renormalization procedure, analogous to the one for the density of states. Taking advantage of the associativity property of the product of matrices, one can group together the product obtained for a given generation in order to find recurrence relations for the next generation.

Suppose that one has obtained the product for chains of generations $n-1$ and $n-2$, then the product for generation n is

$$\left(\prod_{i=1}^{m(n-2)-1} M_i\right) M_C \left(\prod_{j=1}^{m(n-1)-1} M_j\right) = \prod_{k=1}^{m(n)-1} M_k, \qquad (XXII.7)$$

where M_C can be M(short–short) or M(short–long), according to whether the generation is even or odd respectively.

Equation (XXII.7) is evaluated easily for very long chains. In Figure XXII.2 the results are shown for $n=50$ ($m \simeq 10^{10}$), where it is seen that the gap distribution is neatly defined, since no imaginary part is added to E. The fractal structure is evident, if one compares the two spectra shown at different scales in Figures XXII.2a and XXII.2b. The self–similarity is found at the center of the spectrum and at the edges of each cluster. A similar distribution has been reported for the transmission coefficient in a much shorter Fibonacci chain (56 bonds).

A much clearer picture appears when one plots the inverse of the Lyapunov exponent (usually the localization length) against energy. In Figure XXII.3a this plot is shown on a logarithmic scale for the same case as in Figure XXII.2. Notice

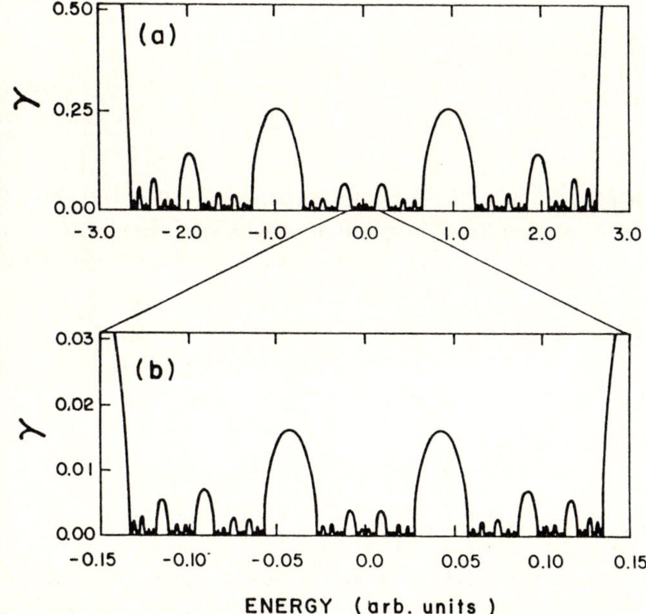

Figure XXII.2. a) Lyapunov exponent $\gamma(E)$ calculated in a chain with $n=50$. b) Magnification of the region around E=0 indicated in a).

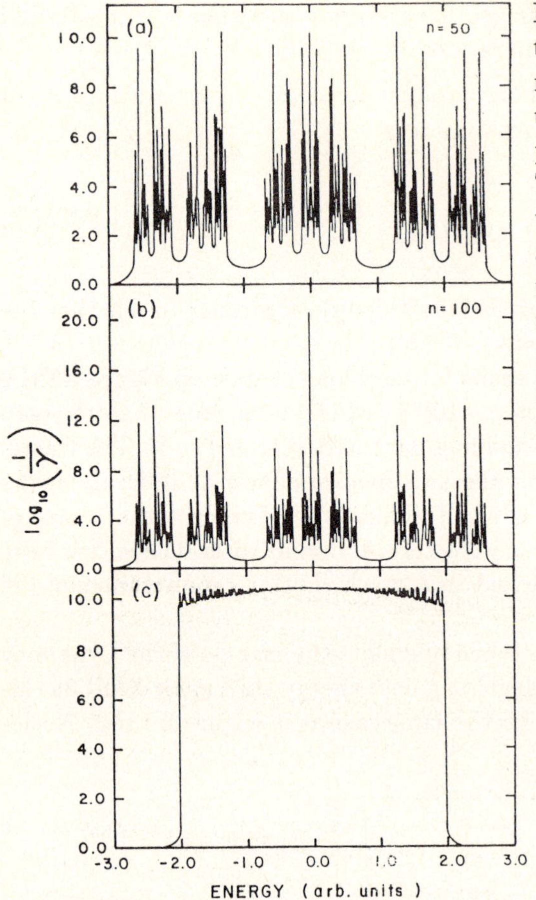

Figure XXII.3 a) Logarithm of the inverse of the Lyapunov exponent (localization length) as a function of energy for a chain with approximately 5.731×10^{10} atoms. b) The same as in a) but for a chain with approximately 2.036×10^{20} atoms. Observe that the localization length of the central peak is of the order of the length of the chain. c) The same quantity calculated in a perfect finite linear chain with the same number of atoms as in a).

that the most localized states, in the middle of the gaps, have an extension of about 7 lattice sites, which is reasonable for surface states. In the limit of a truly infinite chain, without surfaces, there are no accessible states in the gaps attached to infinite perfect chains [XXII.20]. Only the state at E=0 is found to have a localization length of the order of the length of the chain. However this is an artifact because the computer never gives an irrational number. In fact one expects that all the states in the center of a cluster should be extended according to the present criterion.

A most interesting fact is revealed when one compares Figures XXII.3a and XXII.3b, because there is a number of states which conserve their length, independently of the length of the chain. These should be surface states, which in a periodic chain with free ends should only erode the band edges (see Figure XXII.3c for a periodic chain of similar length), but in the Fibonacci spectrum the surface effects are felt everywhere in the spectrum, due to the distribution of bands and gaps. Therefore, if one is looking for a peculiar physical behaviour

200

of quasicrystal, a good choice would be a property in which the interplay of localized and extended states is crucial. An example of such a situation is the Kondo problem, which we briefly discuss below.

XXII.3. Magnetic Interactions through Quasicrystalline Chains

The Kondo effect has been known since the 1930's, when it was discovered that the electrical resistance of simple metals containing magnetic impurities presented a minimum at very low temperatures. Kondo [XXII.21] was the first to explain this puzzle as a many–body effect caused by the magnetic interaction of the localized f or d electrons in the impurities and the conduction electrons in the metal. Since then significant progress has been made in the understanding of this complicated process, and recently Wiegman [XXII.22, XXII.23] has found an exact solution to the Kondo problem using the Bethe anzatz. There are good reviews of the properties of Kondo systems [XXII.24], and amongst them there is the formation of a many–body resonance in the impurity density of states at the Fermi level, known as the Abrikosov–Suhl (AS) peak, at temperatures low compared with the Kondo temperature. This temperature characterizes the crossover from the perturbative interaction of conduction electrons with the localized magnetic moments, to the strong interaction regime, in which the mobile band electrons compensating the localized spins contribute to the formation of the AS resonance. In addition to the single impurity Kondo problem one has to consider the indirect exchange produced by the Ruderman–Kittel–Kasuya–Yosida spin–density oscillations.

In order to investigate the peculiarities of these magnetic interactions in quasicrystals, we have devised a system sketched in Figure XXII.4, in which two magnetic ions with localized d-electrons are connected through a linear chain, which could be a Fibonacci or a periodic system of s-electrons. The

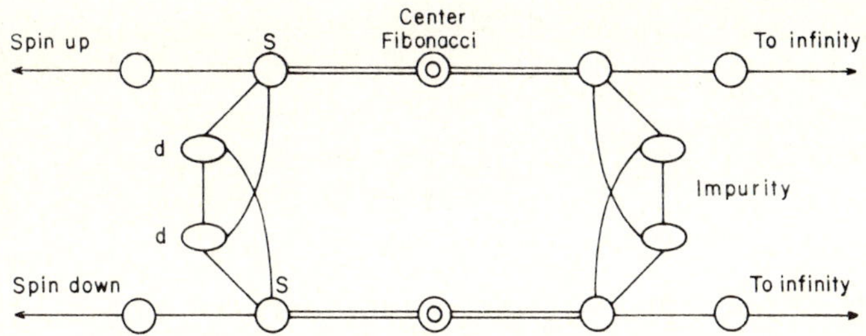

Figure XXII.4. Sketch showing a chain with two magnetic atoms. The apparent ring is due to the fact that the two spin states have been separated in order to show the spin flip terms (curved lines) in the impurities. The atoms in the chain have been renormalized and represented by a double circle. The chains continue to infinity through a Bethe lattice effective medium.

interactions between states are shown by links in the Figure and correspond to
a version of the Hamiltonian introduced by Anderson [XXII.25]:

$$H = E_s \sum_{i\sigma} c^\dagger_{i\sigma} c_{i\sigma} + \sum_{i\sigma} t_{i\sigma} \left(c^\dagger_{i\sigma} c_{i+1\sigma} + c.c. \right) + E_d \sum_{m\sigma} d^\dagger_{m\sigma} d_{m\sigma}$$

$$+ V_{sd} \sum_{m\sigma} \left(d^\dagger_{m\sigma} c_{m\sigma} + c.c. \right) \qquad\qquad (XXII.8)$$

$$+ \frac{J}{2} \sum_m \sum_{\sigma\sigma'} \sum_{\nu\nu'} \overline{\sigma}_{\sigma\sigma'} \cdot \overline{\sigma}_{\nu\nu'} \left(c^\dagger_{m\sigma} d^\dagger_{m\nu} c_{m\sigma'} d_{m\nu'} \right),$$

where the c's create s–electrons and the d's create d–electrons in the impurity
m sites. The Pauli spin matrices obey the relation

$$\overline{\sigma}_{\alpha\beta} \cdot \overline{\sigma}_{\gamma\delta} = 2\delta_{\alpha\delta}\delta_{\beta\gamma} - \delta_{\alpha\beta}\delta_{\gamma\delta}. \qquad\qquad (XXII.9)$$

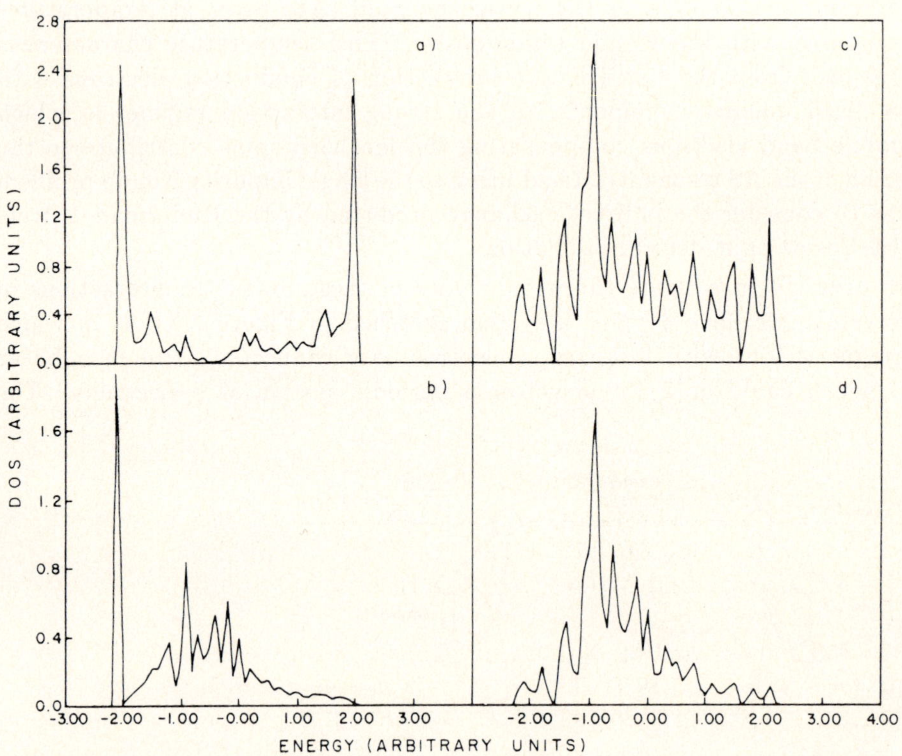

Figure XXII.5. a) Local density of states from the central atom in Fig. XXII.4
in the periodic case $(t_l = t_s)$. The parameters used were $J=2$, $V_{sd}=-1$, $E_d=-0.5$,
$t_l=-1$. b) Local density at the impurity site. c) The same as in a) but for the
Fibonacci case $(t_l=-1$ and $t_s=-1.2)$. d) The same as in b) for the Fibonacci
chain.

Using mean field to linearize (XXII.8) one can solve the problem self–consistently. In Figure XXII.5a the local density of states in the middle site of a pure chain is shown, and in Figure XXII.5b the local density of states at the impurity site shows the Kondo peak nearly at the Fermi level. These should be compared with the corresponding DOS in Figures XXII.5c and XXII.5d for the Fibonacci chain. The number of atoms in the chain is the same in both cases. Notice the effect of the impurity in the central part of the Fibonacci spectrum which is sensed even in the middle of the chain. There is a complicated structure of peaks around the Fermi level in the case of the quasicrystal.

These results are only preliminary and a full discussion of the problem will be published elsewhere. The purpose of presenting them here is only to give an example of the beauty of the physics hidden in quasicrystalline systems.

References

XXII.1 T. C. Choy, Phys. Rev. Lett. **55**, 2915 (1985).

XXII.2 K. Bajema and R. Merlin, Phys. Rev. B **36** 4555 (1987).

XXII.3 R. K. P. Zia and W. J. Dallas, J. Phys. A **18**, L341 (1985).

XXII.4 C. Wang and R. A. Barrio, Phys. Rev. Lett. **61**, 191 (1988).

XXII.5 H. Bohr, Acta Math. **45**, 29 (1924).

XXII.6 A. S. Besicovitch, *Almost Periodic Functions* (Cambrige Press, London, 1932).

XXII.7 M. V. Romerio, J. Math, Phys., **12**, 552 (1971).

XXII.8 J. P. Lu, T. Odagaki and J. L. Birman, Phys. Rev. B **33**, 4809 (1986).

XXII.9 T. Odagaki and L. Friedman, Solid State Commun. **57**, 915 (1986).

XXII.10 Ch. Wang and R.A. Barrio, *Proc. III Int. Conf. on Quasicrystals and Incommensurate Structures* (World Scientific, to be published).

XXII.11 M. Fujita and K. Machida, Solid State Commun. **59**, 61 (1986).

XXII.12 P. Villaseñor-González, F. Mejía-Lira, and J.L. Morán-López, Solid State Commun. **66**, 1127 (1988).

XXII.13 R.A. Barrio and M. del Castillo-Mussott, Solid State Commun. **65**, 775 (1988).

XXII.14 K. Machida and M. Nakano, Phys. Rev. B **34**, 5073 (1986).

XXII.15 N.F. Mott and W.D. Twose, Adv. Phys. **10**, 107 (1961).

XXII.16 F. Deylon and D. Petritis, Commun. Math. Phys. **103**, 441 (1986).

XXII.17 Y. Liu and R. Riklund, Phys. Rev. B **35**, 6034 (1987).

XXII.18 M. Kohmoto, Phys. Rev. B **34**, 5043 (1986).

XXII.19 M. Kohmoto, B. Sutherland, and C. Tang, Phys. Rev. B **35**, 1020 (1987).

XXII.20 J.A. Vergés, L. Brey, E. Louis and C. Tejedor, Phys. Rev. B **35**, 5270 (1987).

XXII.21 J. Kondo, Progr. Theor. Phys., **32**, 37 (1964).
XXII.22 P.B. Wiegman, Pis'ma Zh. Eksp. Teor. Fiz., **31**, 392 (1980).
XXII.23 A.M. Tsvelick and P.B. Wiegman, Adv. Phys., **32**, 453 (1983).
XXII.24 N.B. Brant and V.V. Moshchalkov, Adv. Phys., **33**, 373 (1984).
XXII.25 P.W. Anderson, Phys. Rev., **124**, 41 (1961).

XXIII. Critical Behavior of Quantum Heisenberg Chains

P. Schlottmann and Kong-Ju-Bock Lee

Department of Physics, Temple University, Philadelphia, PA 19122, USA

We critically review the low–temperature behavior of integrable quantum Heisenberg chains, *i.e.*, spin systems for which thermodynamic properties can be numerically or analytically extracted from the thermodynamic Bethe–ansatz equations. In particular we discuss the critical behavior of the spin 1/2 ferromagnet and of its $SU(2)$–invariant extension to arbitrary spin S, the properties of an impurity spin S_{imp} in a spin–one Heisenberg chain and some anomalies of antiferromagnetic chains and the Hubbard model. Logarithmic singularities as a function of field or temperature appear to be a common feature of these isotropic spin chains.

XXIII.1. Introduction

Many properties of one–dimensional quantum spin systems are known. This is specially the case for integrable systems for which the exact diagonalization of the Hamiltonian by means of Bethe's ansatz is possible. The knowledge of the complete set of energy eigenvalues enables one to obtain the free energy and to study the thermodynamic properties of the system. In the thermodynamic limit it is necessary to classify the states and to introduce a density of states for each class of excitations. These densities of states are determined by integral equations (typically an infinite set) known as the thermodynamic Bethe–ansatz equations. The problem now consists of solving this infinite set of integral equations, which is often possible analytically for limiting situations, but requires a numerical effort in the general case.

In the following sections we critically review some of the low–temperature properties of quantum Heisenberg chains obtained so far. In section XXIII.2 we discuss the critical behavior of the isotropic spin–1/2 ferromagnet. These results are extended in section XXIII.3 to the $SU(2)$–invariant Heisenberg chain of arbitrary spin S. The properties of an impurity spin S_{imp} in an otherwise translational invariant spin–one Heisenberg chain are discussed in section XXIII.4. In section XXIII.5 we summarize some anomalous properties of antiferromagnetic chains and the Hubbard model. Conclusions follow in section XXIII.6.

Springer Proceedings in Physics, Vol. 50   **Magnetic Properties of Low-Dimensional Systems II**

Editors: L.M. Falicov · F. Mejía-Lira · J.L. Morán-López © Springer-Verlag Berlin, Heidelberg 1990

XXIII.2. Low–Temperature Behavior of the Spin–1/2 Heisenberg Ferromagnet

The critical behavior of the one–dimensional isotropic spin–1/2 Heisenberg ferromagnet has been studied a variety of methods, which gave quite different results [XXIII.1-XXIII.8] for the critical exponents of the susceptibility γ and of the specific heat α (see Table XXIII.1). This long–standing problem was resolved in part by recent numerical solutions of the coupled thermodynamic Bethe–ansatz integral equations by Schlottmann [XXIII.7] and by Takahashi and Yamada [XXIII.8]. Although the numerical methods employed [XXIII.7,XXIII.8] are considerably different, both calculations agree within error bars and yielded the exponent values $\alpha = -1/2$ and $\gamma = 2$, *i.e.*, the exponents expected from spin waves. A gradual crossover into the critical region is observed with decreasing temperature, which explains the diversity of extrapolated exponents [XXIII.1–XXIII.6] from temperature ranges that were not low enough. Note that the self–consistent one–magnon theory [XXIII.3] necessarily yields the classical spin–wave critical exponents.

We consider the isotropic spin–1/2 Heisenberg chain

$$H = J \sum_{i=1}^{N} \mathbf{S}_i \cdot \mathbf{S}_{i+1} - (1/4)JN - 2H \sum_{i=1}^{N} S_i^z \qquad (XXIII.1)$$

of N sites and periodic boundary conditions in the limit $N \to \infty$. For $J < 0$ the ground state is ferromagnetic with all spins aligned. Flipping a spin one obtains an elementary excitation known as magnon which travels through the chain with real momentum, k. Two or more magnons can form a bound–state. Complex conjugated magnon momenta are associated with bound–states, so that the wave–function falls off exponentially with the relative coordinates, but the center–of–mass travels with real momentum. It is convenient to introduce "rapidities" related to the momenta via $\Lambda = (1/2)\cot(k/2)$, since in the thermodynamic limit the complex rapidities arrange themselves in "strings". A string of rapidities has a common real part (related to the center of mass motion of the bound state), Λ_n, and equally spaced imaginary parts. A bound state of n magnons is then characterized by a string of length n, $n = 1, 2, \ldots$, and a rapidity Λ_n, $-\infty < \Lambda_n < \infty$. Strings with $n = 1$ label the free magnons. The string states can of course not be occupied independently of each other and in thermal equilibrium their population is given by the fugacities $\eta_n(\Lambda)$. The fugacities are determined by an infinite set $(n = 1, 2, \ldots)$ of nonlinear integral equations (derived independently by Takahashi [XXIII.11] and Gaudin [XXIII.12] based on Bethe's famous work [XXIII.13] known as the thermodynamic Bethe–ansatz equations.

An exact analytical solution of this infinite set of nonlinear integral equations appears to be possible only at $T = 0$ (yielding the ferromagnetically ordered ground state and all the excitation energies) and for $T \gg |J|$, the free–spin solution. In the critical region, i.e. at very low T, the solution is an interpolation between the free–spin and zero–temperature limits, namely for small n and Λ the solution is very close to the ferromagnetic one, while if n and/or $|\Lambda|$ are large it asymptotically approaches the free–spin solution. It is then possible to truncate [XXIII.8,XXIII.10] the infinite set of equations at the string–index n_o, replacing the functions for $n > n_o$ by the asymptotical free–spin expression, such that actually only n_o coupled nonlinear integral equations are solved. The procedure is repeated for several n_o (up to $n_o = 280$) and the obtained values for the susceptibility and the specific heat extrapolated to $n_o \to \infty$ (for details see [XXIII.10]). The extrapolated results for χT^2 and $S/T^{1/2}$ are shown in Figs. XXIII.1a and XXIII.1b as a function of $T^{1/2}$. Although the data do not lie on perfect straight lines, the figures show that $\alpha = -1/2$ and $\gamma = 2$ are plausible values of the critical exponents. A more detailed analysis [XXIII.8] yields $\alpha = -0.49 \pm 0.02$ and $\gamma = 2.00 \pm 0.02$. The diverse γ–values listed in Table 1 are in part caused by the small change in curvature of χT^2 around $T \sim 0.02$ $(J = -1)$.

Takahashi and Yamada [XXIII.9], on the other hand, obtained the specific heat and susceptibility of the isotropic ferromagnet by numerically solving the thermodynamic Bethe–ansatz equations of the planar XXZ model and extrapolating the anisotropy to zero $(\Delta \to 1)$. They use the fact that for $\Delta = \cos \pi/n$, $n = 3, 4, \ldots$, the number of integral equations is $n - 1$, i.e., finite. In this way the truncation of an infinite set of equations is eliminated at the cost of the extrapolation $\Delta \to 1$ (the danger of such an procedure is discussed in section XXIII.5). They considered the same temperature range and the same number of integral equations (i.e., n_o) as in [XXIII.8]. The agreement between the two data sets is of the order of or better than 0.8%, i.e. within error bars.

We have argued above that in the critical region the solution interpolates between the $T = 0$ solution (for small n) and the one for $T \gg |J|$ (for large

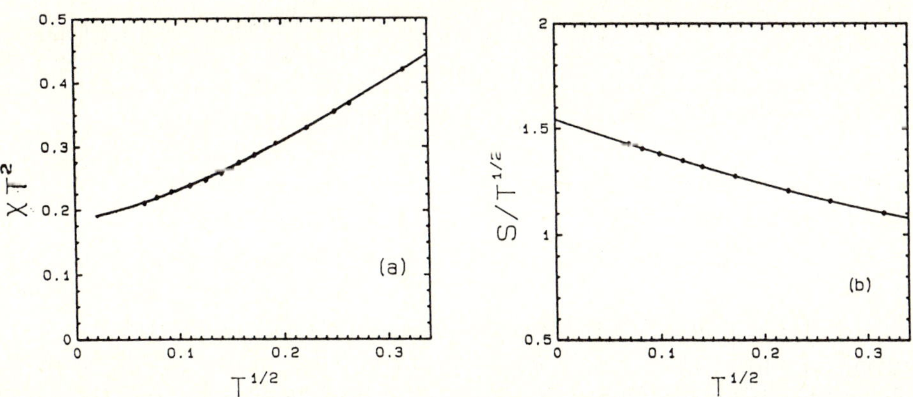

Figure XXIII.1. (a) Susceptibility and (b) entropy of the spin 1/2 Heisenberg ferromagnet as obtained by numerical solution of the Bethe–ansatz equations.

Table XXIII.1. Critical exponents for the Heisenberg ferromagnet

Reference	Method	$-\alpha$	γ
Bonner and Fisher [XXIII.1]	Extrapolation from finite rings (N$\leq$ 11)	0.45–0.50	$\sim$1.80 1.67$\pm$0.07
Baker, Rushbrooke and Gilbert [XXIII.2]	High–temperature series and Padé approximants	—	1.67$\pm$0.07
Kondo and Yamaji [XXIII.3]	Green's functions: self-Consistent one magnon theory	—	$\sim$1.32
Cullen and Landau [XXIII.4]	Monte Carlo	—	$\sim$1.32
Lyklema [XXIII.5,XXIII.6]	Monte Carlo and finite size scaling; $(\gamma - 1)/\nu =$ 1.01$\pm$0.01	0.30$\pm$0.1	1.75$\pm$0.02
Marcu, Müller and Schmatzer [XXIII.7]	Monte Carlo	0.261$\pm$0.013	1.552$\pm$0.008
Schlottmann [XXIII.8]	Numerical solution of Bethe–ansatz equations	0.49$\pm$0.02	2.00$\pm$0.02
Takahashi and Yamada [XXIII.9]	Numerical solution of Bethe–ansatz equations	0.5	2
Schlotmann [XXIII.10]	Analytical solution of Bethe–ansatz equations	0.5	2 with log corrections

n). It is then possible to define a crossover index $n_c(T)$ at which the $T = 0$ and free–spin fugacities are equal:

$$|J|/n_c = T\ln(n_c + 1). \qquad (XXIII.2)$$

For $n_o > n_c$ the solution is then closer to the one of free spins, while for $n < n_c$ the strong coupling ($T = 0$) solution dominates. This means that in thermal equilibrium the average number of correlated spins is $n_c(T)$, i.e., $n_c(T)$ is the correlation length of the system. Solving (XXIII.2) interactively we obtain, for $T \ll |J|$

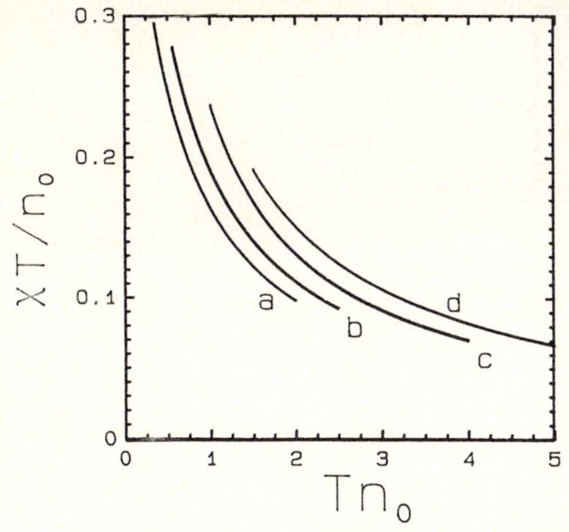

Figure XXIII.2. Finite "string–size" scaling of the susceptibility data for four temperatures, (a) $T/|J| = 0.0045$, (b) 0.0125, (c) 0.03 and (d) 0.05.

$$\xi \sim (|J|/T)\{\mathrm{Ln}^{-1} + \ln(\mathrm{Ln})/\mathrm{Ln}^2 + \ldots\} \qquad (XXIII.3)$$

where $\mathrm{Ln}=\ln(|J|/T)$, corresponding to $\nu = 1$, with logarithmic corrections.

Our numerical results for a given $n_\circ$ are only meaningful if $n_\circ > n_c$, since at most $n_\circ$ spins can be correlated. This is similar although not exactly the same to considering a system of finite size $n_\circ$ larger than the correlation length. The variation of χT as a function of $n_\circ^{-1}$ also resembles a dependence on finite size. Assuming $\nu = 1$ and $\gamma = 2$ we should plot $T\chi(n_\circ,T)/n_\circ$ as a function of $Tn_\circ$ for a finite "string–size" analysis and obtain a universal relation. This is not the case as shown in Fig. XXIII.2 for four isothermals. The curves are almost parallel with a spacing that is roughly logarithmic. This is not totally unexpected since according to (XXIII.2) the relevant variable is $Z = Tn_\circ \ln(n_\circ + 1)$, i.e., the string–size is to be scaled with the correlation length ξ. The plot of $T\chi(n_\circ,T)/n_\circ$ vs. Z is shown in Fig. XXIII.3. All points lie close to one curve; there is some scattering, which is not random, but systematic. The lower–T points lie below the curve, while the high–T data are above. It follows from this universal curve that

$$\chi(T) = (A/T^2)[\mathrm{Ln}^{-1} + \ln(\mathrm{Ln})/\mathrm{Ln}^{-2} + \ldots] \sim \xi/T, \qquad (XXIII.4)$$

i.e., $\gamma = 2$ with logarithmic corrections. Since the dependence of the entropy on $n_\circ$ is much smaller than it is for χ, the leading term of the entropy does not have logarithmic corrections, i.e.

$$S = B(T/|J|)^{1/2} + O(T) \qquad (XXIII.5)$$

with B being close to $(3/2)\times1.042$.

The expressions (XXIII.4) and (XXIII.5) for $\chi(T)$ and $S(T)$ have been extracted analytically from the thermodynamic Bethe–ansatz equations by as-

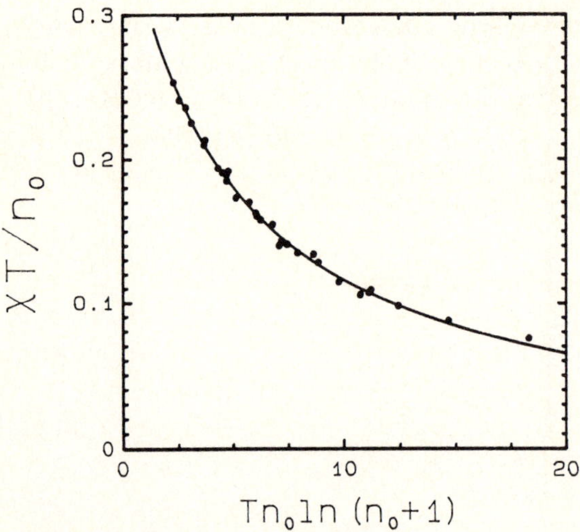

Figure XXIII.3. Finite "string–size" scaling of all the χ–data, now scaled with $Tn_\circ \ln(n_\circ + 1)$, see eq. (XXIII.2).

suming an abrupt crossover between the strong–coupling and free–spin solutions, i.e., string excitations with $n < n_c(T)$ and $|\Lambda|$ smaller than a critical value are suppressed, while all other excitations have the free–spin fugacity. Iteration of the integral equations with this ansatz modifies amplitudes and corrections to scaling, but does not affect the leading temperature dependence. Hence, we must conclude that χ and ξ have logarithmic corrections in the critical region.

XXIII.3. Critical Behavior of the SU(2)–Invariant Ferromagnet of Spin S

The analytical results discussed at the end of the previous section can be extended to arbitrary spin values S and to finite magnetic fields. Unfortunately the standard extension of Heisenberg's model, i.e., model (XXIII.1) with $S > 1/2$, is not integrable. There are, however, integrable generalizations of the Heisenberg chain. Via the quantum inverse scattering method it has been shown [XXIII.14–XXIII.16] that a certain polynomial Q_{2S} of order $2S$ of the quantity $\mathbf{S}_i \cdot \mathbf{S}_{i+1}$ is integrable, i.e.,

$$H_S = J \sum_{i=1}^{N} Q_{2S}\left(\mathbf{S}_i \cdot \mathbf{S}_{i+1}\right) - E_F , \qquad (XXIII.6)$$

where E_F is the energy of the state with all spins aligned. The definition of the polynomials Q_{2S} can be found in [XXIII.14–XXIII.17]. Model (XXIII.6) is the SU(2) generalization of (XXIII.1); for $S = 1/2$ it just reduces to (XXIII.1) and for $S = 1$ it takes the form

$$H_{s=1} = (J/4) \sum_{i=1}^{N} \left\{ (\mathbf{S}_i \cdot \mathbf{S}_{i+1}) - a(\mathbf{S}_i \cdot \mathbf{S}_{i+1})^2 \right\}, \ a = 1. \qquad (XXIII.7)$$

210

Although this model is quite different from the ordinary spin S Heisenberg ferromagnet, its low–temperature behavior is expected to be similar. Thermodynamic Bethe–ansatz equations for this model were derived by Babujian [XXIII.16].

Following the analytic procedure briefly discussed at the end of section XXIII.2 we obtain the critical behavior [XXIII.17] of ξ, χ and C

$$\xi \simeq (|J|/T)\{\mathrm{Ln}^{-1} + \ln(\mathrm{Ln})/\mathrm{Ln}^2 \ldots\}$$

$$\chi \simeq 0.42(2S)^2 (\xi/T), \quad C \simeq 1.6(2ST/|J|)^{1/2}, \qquad\qquad (XXIII.8)$$

where $\mathrm{Ln} = \ln(2S|J|/T)$. A small but finite magnetic field smears the logarithmic corrrections in ξ and χ. In a large magnetic field $(H/T \gg 1)$ we obtain for the free energy of the $S = 1/2$ chain [XXIII.17]

$$F \sim -H - (2/\pi)|J|^{-1/2}T^{3/2} \exp(-3H/T) + \ldots \qquad\qquad (XXIII.9)$$

XXIII.4. Integrable Spin–One Heisenberg Chain with Impurity [XXIII.18]

The $\mathrm{SU}(2)$–invariant spin–chains discussed in the previous section remain integrable if a link to an impurity site is added to the chain, provided the interaction between the impurity spin and its neighboring spins is of a special form. For a chain of spin $1/2$ and an impurity of arbitrary spin S the interaction Hamiltonian was constructed by Andrei and Johannesson [XXIII.19]. We have generalized [XXIII.18] these results to the more interesting situation of a chain of spin 1, i.e., model (XXIII.7) with impurity spin S, by making use of the quantum inverse scattering method. The particular form of the interaction and the thermodynamic Bethe–ansatz equations can be found in [XXIII.18].

We here restrict our discussion to the low temperature properties of the impurity. For ferromagnetic coupling $(J < 0)$ the thermodynamics of the impurity is dominated by the critical behavior of the chain, so that the temperature dependence of χ and C is the same as in (XXIII.8), although the amplitudes are different. The results for antiferromagnetic coupling $(J > 0)$ are more interesting. In the limit $T \to 0$ and for small fields H the free energy yields $(J = +1)$

$$
\begin{aligned}
F &= -\ln 2 - (2/\pi^3)H^2 |\mathrm{Ln}| - (4/\pi^3)H^2 + \ldots, && S = 1/2, \\
F &= -1 - (2/\pi^2)H^2 [1 - \mathrm{Ln}^{-1} - \ln(\mathrm{Ln})/\mathrm{Ln}^2 + \ldots], && S = 1, \\
F &= -1 - 2(S - 1)H - 2SH/\mathrm{Ln} + \ldots && S > 1,
\end{aligned}
$$
$$\qquad\qquad (XXIII.10)$$

where $\mathrm{Ln} = \ln H$. Hence, the susceptibility for $S = 1/2$ diverges logarithmically as $H \to 0$. This result resembles similar ones found for the spin $1/2$ two–channel Kondo problem or quadrupolar Kondo effect [XXIII.20], where the quadrupolar susceptibility diverges logarithmically for a vanishing lattice disortion. This situation is also related to the scattering of conduction electrons on two level systems in metallic glasses [XXIII.21], where now H plays the role of the asymmetry of the tunneling well. For $S = 1$ the impurity is actually no impurity, but just one more site in the chain, so that F represents the bulk free energy. For $S > 1$, on the other hand, the chain spins are too small to compensate the impurity spin, so that the impurity is magnetic with an effective spin $(S - 1)$.

XXIII.5. Anomalous Properties of Antiferromagnetic Heisenberg Chains

In section XXIII.3 we briefly reviewed the SU(2)–extension of the Heisenberg chain to arbitrary spins S and discussed the critical properties for ferromagnetic coupling ($J < 0$). The same model behaves antiferromagnetically if $J > 0$. The most interesting quantity is the $T = 0$ small field susceptibility

$$\chi = (4S/\pi^2)\{1 - S/\mathrm{Ln} - S^2 \ln|\mathrm{Ln}|/\mathrm{Ln}^2 + \ldots\} \qquad (XXIII.11)$$

which shows logarithmic singularities ($\mathrm{Ln} = \ln H$) as $H \to 0$. For $S = 1$ this expression reduces to (XXIII.10) and for $S = 1/2$ to the susceptibility of the ordinary Heisenberg antiferromagnet. The constant zero–field χ was obtained by Griffiths [XXIII.22], the first log–correction by Babujian [XXIII.16] and the second one by us [XXIII.17].

Another possible generalization of the $S = 1/2$ Heisenberg chain is SU($2S + 1$)–invariant. The Hamiltonian just permutes the spin–components of neighboring sites. For spin–1/2 it reduces to (XXIII.1) and if $S = 1$ it corresponds to (XXIII.7) with $a = -1$. The model was solved by nested Bethe–ansätze by Sutherland [XXIII.23]. The $T = 0$ low field susceptibility shows similar logarithmic singularities [XXIII.24] as (XXIII.11)

$$\chi = (4S/\pi^2)\{1 - (N\mathrm{Ln})^{-1} - \ln|\mathrm{Ln}|/(N\mathrm{Ln})^2 + \ldots\}, \qquad (XXIII.12)$$

where $N = 2S + 1$ and $\mathrm{Ln} = \ln H$. For $S = 1/2$ expressions (XXIII.11) and (XXIII.12) are, of course, identical. The same logarithmic fields dependence has been found [XXIII.25] for the degenerate Hubbard model with forbidden site–occupation larger than two.

The nonanalytic field dependence of the susceptibility suggests that at $T = H = 0$ the SU-invariant antiferromagnets behave anomalously. For instance we have shown [XXIII.26] that the $T \to 0$ and $H \to 0$ limits cannot be interchanged, i.e., for $S = 1/2$

$$\lim_{H \to 0} \lim_{T \to 0} C/T = (1 + (e/\pi)^{1/2})/3 \neq \lim_{T \to 0} \lim_{H \to 0} C/T = 2/3. \qquad (XXIII.13)$$

The SU(2) and SU(3) invariant chains for $S = 1$ correspond to $a = \pm 1$; the singular behavior for $|a| = 1$ can be taken as an indication that a gap opens in the excitation spectrum for $|a| < 1$, hence confirming Haldane's conjecture [XXIII.27] that the ordinary $S = 1$ Heisenberg chain ($a = 0$) has such a gap. The logarithmic singularities appear to be associated with an SU–symmetry of the model and cannot be obtained as a limiting process of vanishing anisotropy, e.g., as $\Delta \to -1$ in the $S = 1/2$ XXZ chain. Hence, one cannot get logarithmic corrections in the critical behavior of the $S = 1/2$ ferromagnet using the numerical procedure of [XXIII.9]. This could explain small differences in the numerical results of [XXIII.8,XXIII.9].

XXIII.6. Concluding Remarks

We reviewed the critical behavior of SU(2)–invariant Heisenberg ferromagnets, the properties of an impurity in a Heisenberg chain of spin 1 and some low T small field anomalies of antiferromagnets. All models discussed have SU–symmetry and show logarithmic singularities in the susceptibility as a common feature. These logarithmic corrections are expected to be suppressed if the coupling between neighboring sites is not strictly SU–symmetric.

Finally we would like to discuss the critical exponents of the ferromagnet in the context of scaling and hyperscaling. Since in the ground state all spins are aligned, the spin–spin correlation function is independent of distance, $\eta = 1$, and the magnetization is independent of the field, $\delta = \infty$. The scaling and hyperscaling relations for $T_c = 0$ yield then [XXIII.28] $\gamma = 1 + \nu$ and $\alpha = -\nu$. The first relation is satisfied since $\chi \sim \xi/T$, including the logarithmic corrections. The second relation is not obeyed and led Bonner and Müller [XXIII.29] to conclude that both the scaling and the hyperscaling hypothesis break down for one–dimensional Heisenberg ferromagnets.

We acknowledge the support of the U.S. Department of Energy under the grant DE-FG02-87ER45333.

References

XXIII.1 J. C. Bonner and M. E. Fisher, Phys. Rev. **135**, A640 (1964).

XXIII.2 G. A. Baker, Jr., G.S. Rushbrooke, and H. E. Gilbert, Phys. Rev. **135**, A1272 (1964)

XXIII.3 J. Kondo and K. Yamaji, Prog. Theor. Phys. **47**, 807 (1972).

XXIII.4 J. J. Cullen and D. Landau, Phys. Rev. B **27**, 297 (1983).

XXIII.5 J. W. Lyklema, Phys. Rev. B **27**, 3108 (1983).

XXIII.6 J. W. Lyklema, in *Monte Carlo Methods in Quantum Problems*, ed. M. H. Kalos (Reidel, Dordrecht, 1984), p.145.

XXIII.7 M. Marcu, J. Muller, and K. R. Schmatzer, J. Phys. A **18**, 3189 (1985).

XXIII.8 P. Schlottmann, Phys. Rev. Lett. **54**, 2131 (1985).

XXIII.9 M. Takahashi and M. Yamada, J. Phys. Soc. Jpn. **54**, 2808 (1985); **55**, 2024 (1986).

XXIII.10 P. Schlottmann, Phys. Rev. B **33**, 4880 (1986).

XXIII.11 M. Takahashi, Prog. Theor. Phys. **46**, 401 (1971).

XXIII.12 M. Gaudin, Phys. Rev. Lett. **26**, 1301 (1971).

XXIII.13 H. A. Bethe, Z. Phys. **71**, 205 (1931).

XXIII.14 P. P. Kulish, N. Yu. Reshetekhin, and E. K. Sklyanin, Lett. Mat. Phys. **5**, 393 (1981).

XXIII.15 L. A. Takhtajan, Phys. Lett. **87**A, 479 (1982).

XXIII.16 H. M. Babujian, Nucl. Phys. B **215**, 317 (1983).

XXIII.17 K. Lee and P. Schlottmann, Phys, Rev. B **36**, 466 (1987).

XXIII.18 K. Lee and P. Schlottmann, Phys. Rev. B **37**, 379 (1988).

XXIII.19 N. Andrei and H. Johannesson, Phys. Lett. **100**A, 108 (1984).

XXIII.20 D. L. Cox, Phys. Rev. Lett. **59**, 1240 (1987).

XXIII.21 A. Muramatsu and F. Guinea, Phys. Rev. Lett. **57**, 2337 (1986).

XXIII.22 R. B. Griffiths, Phys. Rev. **133**, A768 (1964).

XXIII.23 B. Sutherland, Phys. Rev. B **12**, 3795 (1975).

XXIII.24 Y. Jee, K. Lee, and P. Schlottmann, unpublished.

XXIII.25 K. Lee and P. Schlottmann, unpublished.

XXIII.26 K. Lee and P. Schlottmann, J. Phys. Cond. Matt., in press.

XXIII.27 F. D. M. Haldane, Phys. Rev. Lett. **50**, 1153 (1983).

XXIII.28 G. A. Baker and J. C. Bonner, Phys. Rev. B **12**, 3741 (1975).

XXIII.29 J. C. Bonner and G. Muller, Phys. Rev. Lett. **56**, 1617 (1986).

XXIV. Long-Range Order in One-Dimensional Quasiperiodic Magnetic Chains

I.I. Satija[1], F. Nori[2], and M.M. Doria[3]

[1]Department of Physics, George Mason University, Fairfax, VA 22030, USA
[2]Institute for Theoretical Physics, University of California,
  Santa Barbara, CA 93106, USA
[3]Instituto de Fisica, C.P. 20516, Universidade de São Paulo,
  São Paulo, 01498 SP, Brazil

We study the one dimensional quantum Ising model in a transverse magnetic field where the exchange couplings are ordered according to a quasiperiodic sequence. Three cases of sequences are considered; the Fibonacci, the Thue–Morse and the modulated cosine. They all undergo the phase transition signaled by the onset of long–range order in the magnetization. However they display novel critical properties —not present in the periodic case— concerning the scaling behavior of the energy spectrum.

XXIV.1. Introduction

Epitaxial growth techniques have allowed the production of new magnetic structures by artificially depositing rare–earth base materials on a crystalline substrate [XXIV.1]. Merlin [XXIV.2] *et al.* pioneered the use of aperiodicity in nonmagnetic superlattices. More recently, quasiperiodic (QP) ordering of layers and multilayers of magnetic materials have produced very complicated spin arrangements [XXIV.3]. Quasiperiodic sequences are neither periodic or random thus offering situations in between these two extreme cases. Their quasiperiodicity implies fractal properties on the system [XXIV.4]. For the *one–dimensional quantum* spin model studied here this means that the energy spectrum forms a Cantor Set.

The quantum Ising model in a transverse magnetic field with two distinct nearest–neighbor spin couplings, $J(n)$, is up to a similarity transformation given by

$$H = - \sum_{n=1}^{N} J(n)\sigma^x(n)\sigma^x(n+1) + \sigma^z(n) \ . \qquad (XXIV.1)$$

The $\sigma^a(n)$ are the Pauli matrices associated to the site n and the transverse magnetic field has been normalized to one. The existence of the duality transformation $\sigma^x(n).\sigma^x(n+1) \rightarrow \mu^z(n)$ and $\sigma^z(n) \rightarrow \mu^x(n-1).\mu^x(n)$, where $\mu^a(n)$ is a new set of Pauli matrices, ensures that having the sequence $J(n)$ on the magnetic field instead will render the same properties in the thermodynamic limit. In our numerical study on finite systems we treat the model with periodic boundary conditions. The three distinct cases listed below were studied.

Springer Proceedings in Physics, Vol. 50   **Magnetic Properties of Low-Dimensional Systems II**
Editors: L.M. Falicov · F. Mejía-Lira · J.L. Morán-López © Springer-Verlag Berlin, Heidelberg 1990

(FB) Fibonacci. The size of the finite system is taken as a Fibonacci number $N = F_N$, which are defined through the rule $F_1 = F_2 = 1$, $F_N = F_{N-1} + F_{N-2}$. The number $\sigma_N = F_N/F_{N-1}$ is called a rational approximant to the golden mean since it approaches the value $\sigma = (\sqrt{5}+1)/2$ for increasing N. Consider the sequence defined by $value(n) = 1 + [n/\sigma_N]/\sigma_N - [(n-1)/\sigma_N]/\sigma_N$ where the symbol $[X]$ stands for the largest integer contained in X. The sequence has periodicity F_N and only contains two distinct elements. Depending if $value(n)$ is 1 or $1 + 1/\sigma_N$ in a site n, we associate the coupling λ or $r.\lambda$ respectively to the sequence $J(n)$.

(TM) Thue–Morse. The Thue–Morse sequence of order N has $N = 2^M$ elements composed of two symbols, 0 and 1, defined recursively as follows, $\epsilon_0 = 0$, $\epsilon_{2n} = \epsilon_n$, $\epsilon_{2n+1} = \bar{\epsilon}_n$ Hence in a finite chain of N sites the couplings λ or $r.\lambda$ are associated to the site in correspondence with the Thue–Morse sequence.

(MC) Modulated Cosine. Here the exchange interaction for a finite chain of size F_N is given by $J(n) = \lambda \cos 2\pi n/\sigma_N$.

The coupling r for the FB and TM models is kept fixed and characterizes the degree of quasiperiodicity of the system since for $r = 1$ periodicity is recovered. For the three models varying λ takes the system through a phase transition that happens at a critical value λ_C very much like in the periodic case [XXIV.5]. Qualitatively one can see this transition in the following way. When the coupling among spins is weak, the aperiodic quantum spin chain does not exhibit long range order in the x–direction. For large values of the inter–spin coupling energy, long–range order is obviously present in this x–direction. This transition exists in the above models similarly to the periodic case.

We recall [XXIV.6] that the zero temperature study of model (1) describes the phase diagram of a *two–dimensional classical* Ising model at finite temperature having quasiperiodicity in one of the directions [XXIV.7]. The Hamiltonian (XXIV.1) is the τ continuum limit of this classical Ising model and the coupling constant λ corresponds to the inverse of temperature.

The major point of this work is that all quasiperiodic models studied so far show a phase transition to long–range order signaled in the electronic states in a way not present in the periodic model. Here we will describe the evidence supporting the *non–analyticity* of the scaling exponent δ at criticality for the FB [XXIV.7] and TM [XXIV.8] sequences. The exponent δ describes the scaling of the total band width with the size of the system. Let TBW represent the sum of all energy bands in a chain with N sites with periodic boundary conditions. Then

$$TBW \propto \frac{1}{N^\delta} \ . \qquad\qquad (XXIV.2)$$

For a *localized* spectrum no scaling in the band–widths can be performed since bands do not exist even for reasonably small periodic chains. One cannot define this exponent δ for such situation. The opposite situation is the *extended* spectrum where enlarging the chain size does not alter the width of the energy bands ($\delta = 0$). Values of delta other than zero describe a *critical* spectrum. For the MC potential we found strong evidence [XXIV.9] of a phase transition in the *nature* of the energy spectrum that goes from *critical* to *localized* at this critical coupling λ_C. This *electronic transition* does not take place for the FB and TM which display a *critical* spectrum at both sides of the transition.

XXIV.2. The Model

In order to study Hamiltonian (XXIV.1) we express it, using the Jordan–Wigner transformation, as a model in terms of fermionic fields,

$$H = c^\dagger A c + (c^\dagger B c^\dagger + h.c.)/2 \qquad (XXIV.3)$$

where $c = (c_1, c_2, \ldots, c_N)$, and c_n are anticommuting fermionic operators associated to the sites n. The nonzero elements of the matrices A $(A^t = A)$ and B $(B^t = -B)$ are defined by

$$A_{n,n+1} = -J(n), \; A_{1,N} = -J(N), \; A_{n,n} = -2, \qquad (XXIV.4)$$

$$B_{n,n+1} = -J(n), \; B_{1,N} = J(N) . \qquad (XXIV.5)$$

The usual Bogoliubov transformation does not work here because of the aperiodicity (i.e., the states with momenta k and $-k$ are not the only ones which have nonzero matrix elements). Therefore in order to diagonalize this Hamiltonian one needs the method of Lieb, Schultz and Mattis [XXIV.10] which requires the diagonalization of the matrix $D = (A + B)(A - B)$. This matrix defines a generalized tight–binding equation given by

$$J(n)\psi_{n-1} + (1 + J(n)^2)\psi_n + J(n+1)\psi_{n-1} = \frac{E^2}{4}\psi_n \qquad (XXIV.6)$$

ψ_n describes an associated free fermion at the site n.

XXIV.3. The Magnetic Transition

The phase transition characterizing the onset of magnetic long–range order takes place at a λ_C whose value must satisfy [XXIV.11] the equation

$$\prod_{n=1}^{N} J(n) = 1. \qquad (XXIV.7)$$

The following values are obtained: $\lambda_C(r) = r^{-(\sigma_N)^{-2}}$ and $\lambda_C(r) = 1/\sqrt{r}$ for finite FB and TM sequences respectively. $\lambda_C = 2$ for sufficiently large chains for the MC sequence. One numerically verifies that at this critical coupling the energy gap, that is the difference between the first–excited and ground–state energies, vanishes as $|\lambda - \lambda_C|$ like in the periodic case ($\nu = 1$). The onset of

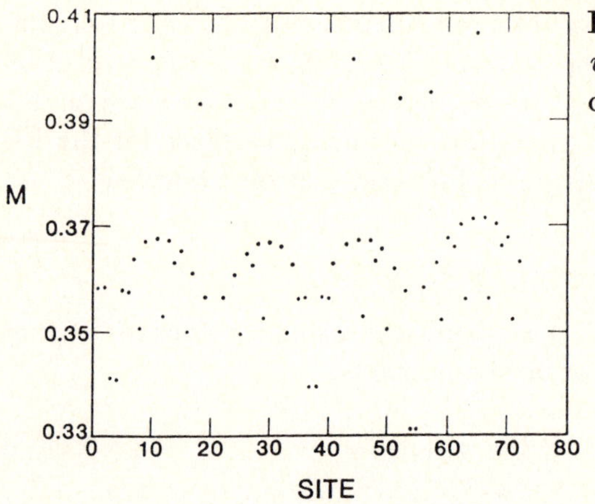

Figure XXIV.1. The magnetization *vs.* site for the FB sequence at criticality for a chain of 144 sites, r=0.5.

long–range order is characterized by the spin correlation in the x direction,

$$\rho(n,\lambda) = \langle G, \sigma^x(n)\sigma^x(n + \frac{F_N}{2})G\rangle, \qquad (XXIV.8)$$

where G is the ground state of (XXIV.1).

For the corresponding two–dimensional classical Ising model this quantity describes the average site magnetization [XXIV.6].

The critical exponent happens to be the same of the periodic case $\beta = 1/8$. Thus one concludes that the lack of translational invariance causes no change in universality class for these models. The novelty for the quasiperiodicity models is that the magnetization becomes site dependent, contrary to the periodic case (see Fig. XXIV.1).

XXIV.4. The Energy–Spectrum Transition

For an infinite chain periodic every N sites, the energy spectrum consists of N bands and N–1 gaps. Increasing values of N render the width of each band narrower. So in this limit the sum of all band widths TBW, has measure zero and scales with the size of the system as $TBW \propto N^{-\delta}$. Our numerical results [XXIV.7,XXIV.8] give the following linear dependence for the FB and TM sequences close to the critical coupling:

$$\delta(\lambda) - \delta(\lambda_c) = b_-(\lambda - \lambda_c), \ \lambda > \lambda_c, \qquad (XXIV.9)$$

$$\delta(\lambda) - \delta(\lambda_c) = b_+(\lambda_c - \lambda), \ \lambda_c > \lambda, \qquad (XXIV.10)$$

where $b_\pm$ are positive functions of r. The exponent δ for the TM is larger than the corresponding FB case (see Fig. XXIV.2), thus indicating that the TM sequence is more disordered. The linear behavior found in both sides of the

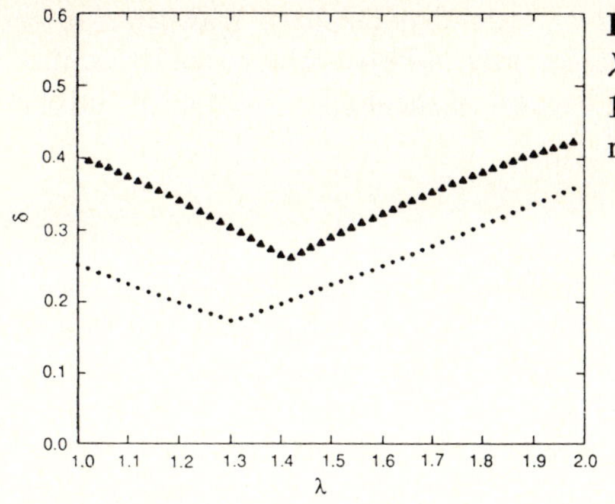

Figure XXIV.2. The δ exponent *vs.* λ plotted for both the TM (triangles, 128 sites) and FB (circles,144 sites) models, r=0.5.

transition leads us to conjecture the *non–analiticity at criticality* for such models. For the MC sequence [XXIV.9] we found that for $\lambda < \lambda_c$ the total band width, although finite for a finite size cyclic chain, it goes to zero as $N^{-\delta}$. A very mild dependence of δ on λ was found probably due to finite size effects ($\lambda \approx 1.0\pm0.1$). For $\lambda > \lambda_c$, we found negligible values of the total band width meaning that δ diverges to infinity (see Fig. XXIV.3). Apart from looking at the total band width, the nature of the states both above and below λ_c was also confirmed by looking at the scaling properties of various individual states. Individual band widths are found to decay algebrically for $\lambda < \lambda_c$ and exhibit exponential decay above the critical coupling. Hence we conjecture that a transition to localization is happening at the same critical coupling where the onset of long–range order takes place. Notice that the presence of one transition does not necessarily imply the other. The FB and TM sequences do have long–range order transition with no transition in the *nature* of the electronic states. From the other side

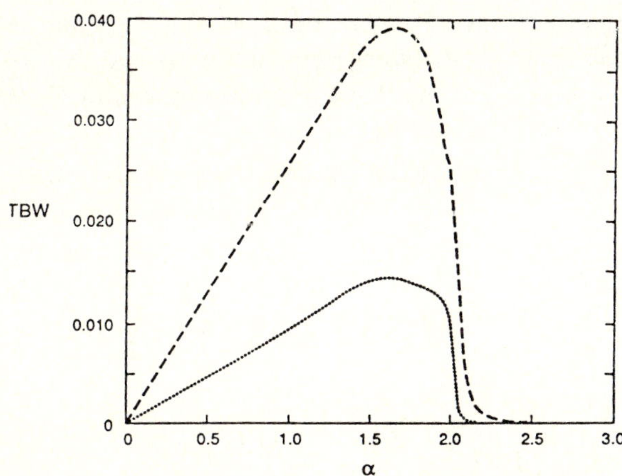

Figure XXIV.3. The total band width *vs.* λ for the modulated cosine (233 sites, dotted line) and (89 sites, dashed lines).

the Harper's equation [XXIV.12] has an extended–localized transition in the energy spectrum without any long–range order. This is the so–called Aubry–Andre [XXIV.12] transition occurring because of the dual properties of Harper's equation.

XXIV.5. Conclusion

In this work we tried to describe some of the results found in our investigation of quasiperiodic magnetic models undergoing a phase transition. We find that the magnetic transition, which signals the onset of long–range order, exists in the quasiperiodic models within the same universality class of the periodic model. However it adds two new features: (1) a modulating magnetization as a function of position along the chain, and (2) additional features in the electronic states. For the FB and TM sequences this means the nonanaliticity of the delta exponent at criticality as well as a linear behavior around λ_C at the two sides of the transition. For the MC we found strong evidence for a transition from critical to localized states occurring simultaneously at the same λ_C. For further information concerning the quasiperiodic XY model [XXIV.13] and characterization of the Cantor spectrum through $F(\alpha)$ curve we refer the reader to previously published material.

This work was done at the Center For Nonlinear Studies, Los Alamos National Laboratory under a partial support of the U. S. Dept. of Energy, contract W–7405–ENG–36.

References

XXIV.1 M. B. Salamon, S. Sinha, J. J. Rhyne, J. E. Cunningham, R. Erwin and C. P. Flynn, Phys. Rev. Lett. **56**, 259 (1986); J. Kwo, E. M. Gyorgy, D. B. McWhan, M. Hong, F. J. DiSalvo, C. Vettier, and J. E. Bower, *ibid.* **55**, 1402 (1985); R. W. Erwin, J. J. Rhyne, M. B. Salamon, J. Borchers, S. Sinha, R. Du, J. E. Cunningham, and C. P. Flynn, Phys. Rev. B **35**, 6808 (1987).

XXIV.2 R. Merlin, K. Bajema, R. Clarke, F.-Y. Juang, and P. K. Bhattacharya, Phys. Rev. Lett. **55**, 1768 (1985).

XXIV.3 C. Vettier, D. B. McWhan, E. M. Gyorgy, J. Kwo, B. M. Buntschuh, and B. W. Batterman, Phys. Rev. Lett. **56**, 757 (1986); C.F. Majkrazac, J. W. Cable, J. Kwo, M. Hong, D. B. McWhan, Y. Yafet, J. V. Waszczak, and C. Vettier, Phys. Rev. Lett. **56**, 2700 (1986).

XXIV.4 M. Azbel, Zh. Eksp. Teor. Fiz **46**, 929 (1964) [Sov. Phys. JETP **19**, (1964)]; D.R. Hofstader, Phys. Rev. B **14**, 2239 (1976).

XXIV.5 P. Pfeuty, Ann. Phys. (N.Y.) **57**, 79 (1970).

XXIV.6 E. Fradkin and L. Susskind, Phys. Rev. D **17**, 2637 (1978).

XXIV.7 M. Doria and I. Satija, Phys. Rev. Lett. **60**, 444, (1988).

XXIV.8 M. Doria, F. Nori, and I. Satija, Phys. Rev. B **39**, 6802 (1989).

 XXIV.9 I. Satija and M. Doria, submitted to Phys. Rev. Lett.

XXIV.10 E. Lieb, T. Schultz, and D. Mattis, Ann. Phys. (N. Y.) **16**, 407 (1961).

XXIV.11 P. Pfeuty, Phys. Lett. **72**A, 245 (1979).

XXIV.12 P. G. Harper, Proc. Phys. Soc. London Sect. A **68**, 875 (1955); S. Aubry and G. Andre, Proc. of the Israel Physical Society, edited by C. G. Kuper (Hilger, Bristol, 1975) Vol 3, 133.

XXIV.13 I. Satija and M. Doria, Phys. Rev. B. **38**, 5174 (1988).

XXV. Magnetic Properties of Small Transition-Metal Clusters

J. Dorantes-Dávila, G.M. Pastor, and K.H. Bennemann*

Freie Universität Berlin, Institut für Theoretische Physik,
Arnimallee 14, D-1000 Berlin 33, Germany

Results for the magnetic moment $\bar{\mu}_n$ and spin–order of small clusters of Fe_n, Cr_n, and Ni_n are presented. The calculations were performed using a tight–binding Hubbard Hamiltonian in the unrestricted Hartree–Fock approximation. In particular we obtain that $\bar{\mu}_n > \mu_b$, and that moments and spin–order depend sensitively on cluster structure. For Fe_n *bcc*–like structures order ferromagnetically, as *fcc*–like Ni_n does, and Cr_n *bcc*–like orders antiferromagnetically. Regarding the critical behavior for $T \rightarrow T_C$, deviation from bulk occurs at a temperature T^* with $(T^*\text{-}T_C) \sim R_n^{1/\tau}$.

Clearly, the magnetic properties of small transition–metal clusters M_n are of great interest, because of their potential for technical applications. Furthermore, small clusters offer a new possibility for studying surface magnetism [XXV.1]. In general one expects for increasing cluster size, that Heisenberg type local moment behavior changes into itinerant magnetism. Due to the interplay of coordination number and interatomic distance (cluster relaxation) the molecular fields, short–range spin–order, etc. are different in small clusters than in bulk solids. Thus, one expects an atomic–environment dependence of the magnetic moments $\mu_n(i)$ (where i refers to the atomic shell of the cluster). Since long–wavelength fluctuations, $q \rightarrow 0$, are cut–off at $q^{-1} \leq R_n$ (where R_n refers to the radii of the clusters assumed to be spherical), the specific heat $C_V(T)$, the susceptibility $\chi_T(q)$, etc. diverge less strongly or are rounded off for $T \rightarrow T_C$. Clusters with $R_n \leq \xi_{sr}$, where ξ_{sr} is the short–range spin order correlation–length, are expected to be magnetic even at $T > T_C$, and also deviations from bulk critical behavior are expected if $\xi(T) \geq R_n$.

Experimentally not much is known at present about the magnetic properties of small transition–metal clusters such as Fe_n, Cr_n, Ni_n. Recently Cox *et al.* [XXV.2] performed magnetic deflection experiments on free Fe–clusters. They (observed) concluded $\bar{\mu}_n > \mu_b$ ($n \leq 17$) and strong size dependence (oscillatory?) of $\bar{\mu}_n$, the average magnetic moment. Furthermore, for small γ–Fe particles [XXV.3] spin order, similar to the one found in bulk γ–Fe, is observed.

In the following we present results for $\bar{\mu}_n$, the local magnetic moments $\mu_n(i)$, and spin–order which is different for *bcc*, *fcc*, or icosahedral cluster structure. Also results are presented for the directional and magnitude spin–fluctuations as

---

* On leave of absence from Departamento de Física, CINVESTAV–IPN, Apdo. Postal 629, 07000 México D.F.

Springer Proceedings in Physics, Vol. 50   **Magnetic Properties of Low-Dimensional Systems II**
Editors: L.M. Falicov · F. Mejía-Lira · J.L. Morán-López © Springer-Verlag Berlin, Heidelberg 1990

T→T$_C$. These results were obtained by using a tight–binding Hubbard Hamiltonian in the unrestricted Hartee–Fock approximation. Such a theoretical method has been successfully used for bulk and surface magnetism and allows us, in particular, to study the interdependence of structure and magnetism and to determine how magnetism stabilizes the cluster structure (*bcc vs. fcc* Fe$_n$, for example). Mostly we neglect *s*–electrons and size dependence of *s–d* hybridization, etc. We have reasons to believe that the cluster size dependence of most transition–metal cluster properties results dominantly from *d*–electrons.

Thus, the Hubbard Hamiltonian for the *d*–electrons is given by

$$H = \sum_{i\sigma} \epsilon_i n_{i\sigma} + \sum_{ij} t_{ij}^{\alpha\beta} d_{i\alpha\sigma}^{+} d_{j\beta\sigma} - E_{dc}, \qquad (XXV.1)$$

with

$$\epsilon_{i\sigma} = \epsilon_p^0 + U\Delta n(i) - \sigma\frac{J}{2}\mu(i) \ . \qquad (XXV.2).$$

Here, $\epsilon_{i\sigma}$ is the energy of a *d*–electron with spin σ at site i, U and J refer to the interatomic (direct) Coulomb interaction and the exchange interaction, respectively, $t_{ij}^{\alpha\beta}$ denotes the hopping integral, and

$$E_{dc} = \frac{1}{2} \sum_{i\sigma} (\epsilon_{i\sigma} - \epsilon_d^0) n_{i\sigma} \qquad (XXV.3)$$

corrects for double counting. One obtains from the Hamiltonian (XXV.1) for the local magnetic moments

$$\mu_n(i) = n_{i\uparrow} - n_{i\downarrow}, \qquad (XXV.4)$$

and

$$n(i) = n_{i\uparrow} + n_{i\downarrow}, \qquad (XXV.5)$$

where

$$n_{i\sigma} = \int_{-\infty}^{\epsilon_F} d\epsilon N_{i\sigma}(\epsilon). \qquad (XXV.6)$$

Here, $N_{i\sigma}(\epsilon)$ is the total electronic density of states at site i and is determined as usually from the Hamiltonian (XXV.1) with the help of the recursion technique for the electronic Green function. Furthermore, the average magnetic moment is defined by

$$\overline{\mu}_n = \frac{1}{n} \sum_i \mu_n(i). \qquad (XXV.7)$$

To determine the cluster structure, bond–length relaxation, etc., we add to the cohesion energy E$_{coh}$ resulting from the Hamiltonian, the Born–Mayer potential,

$$E_R = \frac{1}{2n} \sum_i z_i A e^{-p\left(\frac{d}{d_b}\right)-1} . \qquad\qquad (XXV.8)$$

Here z_i is the coordination number, d the bond length (b=bulk), and p and A constants fitted to bulk properties.

For the calculations [XXV.4] we use the parameters n_d=5,7, and 9, bandwidth W_d=7,6, and 5 eV, J=0.56, 0.73, and 0.50 eV, U=6.3, 5.4, and 4.5 eV and μ_b=0.6, 2.21, and 0.60 μ_B , for Cr_n , Fe_n , and Ni_n , respectively.

The results presented in Table XXV.1, may be summarized as follows:

(a) $\overline{\mu}_n > \mu_b$, in particular for small Cr–clusters, due to $z_{eff} < z_b$, $W < W_b$.

(b) $\mu_n(i)$ depends considerably on atomic position i and cluster structure (as well as $\overline{\mu}_n$), bond–length d (cluster relaxation). For Fe and Cr $\mu_n(i)$ increases as i=0→1,2, etc. This is not the case for Ni_n because of electron charge transfer.

(c) For $n > 15$, approximately, properties begin to converge towards bulk properties.

(d) Spin–order: bcc–like Fe_n is ferromagnetic, fcc–like Fe_n shows antiferromagnetism like γ–Fe with smaller magnetic moments (for example Fe_{13}: ↑↓↑). Magnetic energy (gain for $n = 13$ is 0.3 eV) stabilizes bcc Fe_n . Body–centered cubic like Cr_n is antiferromagnetic, $\mu_n(i) > \mu_b$ at the cluster surface (surface magnetism of Cr), smaller $N_{i\sigma}(\epsilon_F)$ and larger gap Δ_n between bonding and antibonding states than in bulk Cr. Face–centered cubic Ni_n clusters order ferromagnetically, and for example $\overline{\mu}_{13}$=0.96$\mu_B > \mu_b$. Magnetic energy gain stabilizes fcc Ni_n vs. icosahedral Ni_n (for about 0.07 eV).

(e) The trend is towards more itinerancy as the cluster size increases (as obtained from $N_{i\sigma}$, $\mu_n(i)$, molecular fields ξ, etc.).

(f) Stronger short–range order, $T_{sr}(n) > T_{sr}(b)$, in small clusters, mainly due to $\mu_n(i) > \mu_b$. Here T_{sr} is the temperature at which short–range spin correlation is destroyed.

The oscillatory behavior found for $\overline{\mu}_n$ of Fe_n results from the interplay of change of coordination number and bond length (see Figure XXV.1). Including s–electrons, which are expected to increase the bond–length relaxation in particular for small clusters, might change our results for the smaller clusters. It remains to be seen whether the strong variation of $\overline{\mu}$ is real. Note, the experimental results imply $\overline{\mu}$ only indirectly. For $n = 9$ Jahn–Teller distortion may change our value of $\overline{\mu}_n$.

The results shown for $N_{i\sigma}(\epsilon)$ (see Figure XXV.2) and $n = 15$ may serve as a test for the accuracy of our calculations. Comparison with LSDF–calculations indicates that our numerical method produces reliable results. Neglecting the size dependence of s–d hybridization might be most delicate in the case of Ni_n .

The trends towards more itinerancy as the cluster size increases might be inferred from the results obtained for $\Delta E_n(\xi, T)$ of small Fe_n (see Figure XXV.3).

Here, ξ denotes the local molecular field acting on a particular magnetic moment [XXV.1].

Table XXV.1. Size–dependence of various properties of Fe_n, and Cr_n. The cluster structure is assumed. 1=1, 2, etc. refers to the central cluster atom, first–shell atoms, etc. For comparison some $X\alpha$ results are given. Spin–order is indicated.

	Str.	$E_{coh}(n)$	d_n/d_b	$\bar{\mu}$	$\mu(1)$	$\mu(2)$	$\mu(3)$	$\mu(4)$	$\mu(5)$	
Fe_9	bcc	1.33	0.91	2.31	0.40	2.55				
		(1.05)			(2.98)	(2.91)	(2.98)	($\ldots$)=unrelaxed		
Fe_{11}	bcc	1.36	0.98	2.96	2.85	2.98	2.93			
		(1.33)			(2.97)	(2.91)	(2.99)	(2.94)		
Fe_{13}	bcc	1.61	0.97	2.52	0.37	2.63	2.82			
		(1.55)			(2.96)	(2.89)	(2.98)	(2.94)		
Fe_{13}	fcc	1.42	0.98	1.94	−1.67	2.24		↑↓↑ ($\mu(i) > \mu(\gamma$–Fe$)$)		
		(1.40)			(2.04)	(−1.74)	(2.36)			
Fe_{13}	icos.	1.41	0.98	2.17	−1.95	2.51				
		(1.39)			(2.27)	(−2.0)	(2.63)			
Fe_{15}	bcc	1.62	0.97	2.58	0.48	2.64	2.85	↑↑↑↑↑		
		(1.58)			(2.72)	(1.23)	(2.80)	(2.70)		
		SCF–$X\alpha$:			(0.9)	(2.8)	(2.8)			
Fe_{19}	fcc	1.42	0.99	1.86	−1.07	1.81	2.48	↑↑↓↑↑		
		(1.41)			(1.91)	(−1.29)	(1.88)	(2.52)		
Fe_{27}	bcc	(1.39)			(2.81)	(2.86)	(2.62)	(2.82)	(2.93)	
Fe_{43}	fcc	(1.69)			(1.22)	(−1.37)	(−0.90)	(0.87)	(2.48)	
Fe_{51}	bcc	(1.72)			(2.43)	(1.27)	(1.86)	(1.55)	(2.62)	(2.81)
bulk	bcc	2.0	1.0	2.21	2.21					
Cr_8	sc	0.75	0.96	2.22	2.22					
		(0.63)			(2.22)	(2.22)				
Cr_9	bcc	1.08	1.01	3.87	−2.70	4.69		↑↓↑		
		(1.07)			(3.86)	(−2.63)	(4.67)			
Cr_{15}	bcc	1.78	0.98	0.32	−1.74	2.88	−2.74	↓↑↓↑↓		
		(1.75)			(0.33)	(−2.13)	(3.23)	(−3.13)		
		SCF–$X\alpha$:		(−0.7)	(4.1)	(−3.4)				
Cr_{27}	bcc	(1.61)			(1.98)	(1.89)	(−2.36)	(3.32)	(4.20)	
Cr_{51}	bcc	(2.02)			(1.24)	(−1.58)	(2.20)	(−2.11)	(−2.55)	(3.77)
bulk	bcc	3.32	1.0	0.0	0.60	−0.60				

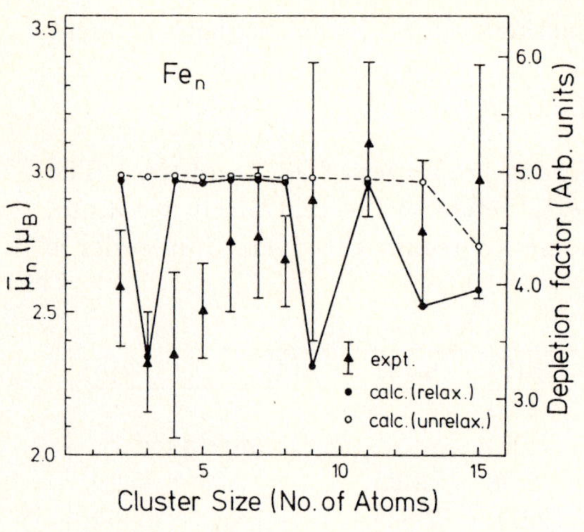

Figure XXV.1. Size dependence of $\bar{\mu}_n$ for Fe$_n$. For comparison experimental results by Cox *et al.* [XXV.2] are shown. $\bar{\mu}_n^{\ exp}$ is derived from the depletion factor of Zeeman experiments.

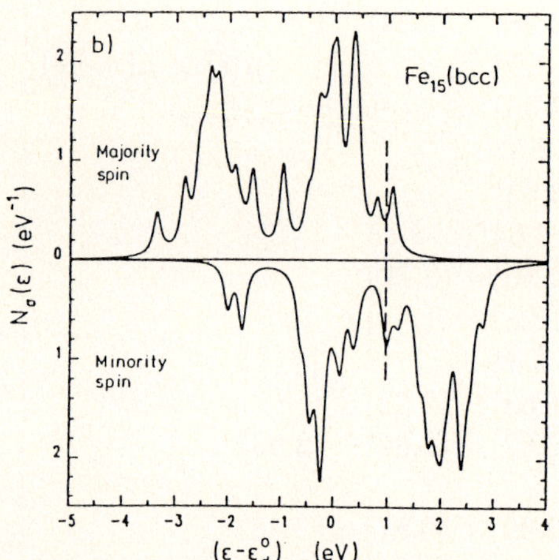

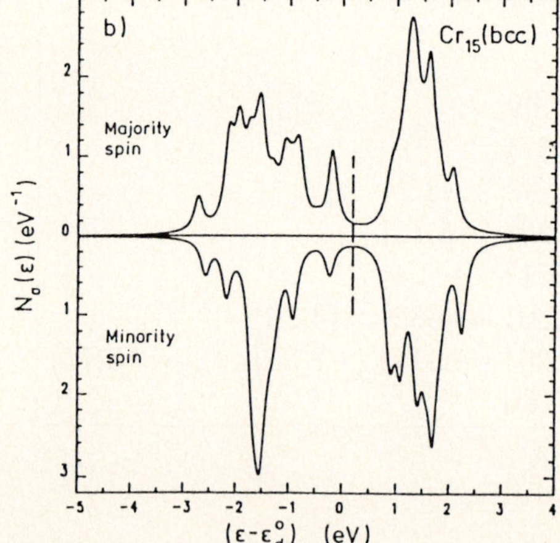

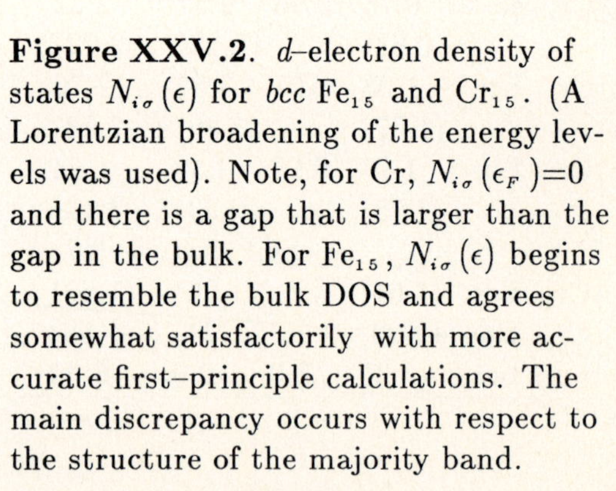

Figure XXV.2. *d*–electron density of states $N_{i\sigma}(\epsilon)$ for *bcc* Fe$_{15}$ and Cr$_{15}$. (A Lorentzian broadening of the energy levels was used). Note, for Cr, $N_{i\sigma}(\epsilon_F)=0$ and there is a gap that is larger than the gap in the bulk. For Fe$_{15}$, $N_{i\sigma}(\epsilon)$ begins to resemble the bulk DOS and agrees somewhat satisfactorily with more accurate first–principle calculations. The main discrepancy occurs with respect to the structure of the majority band.

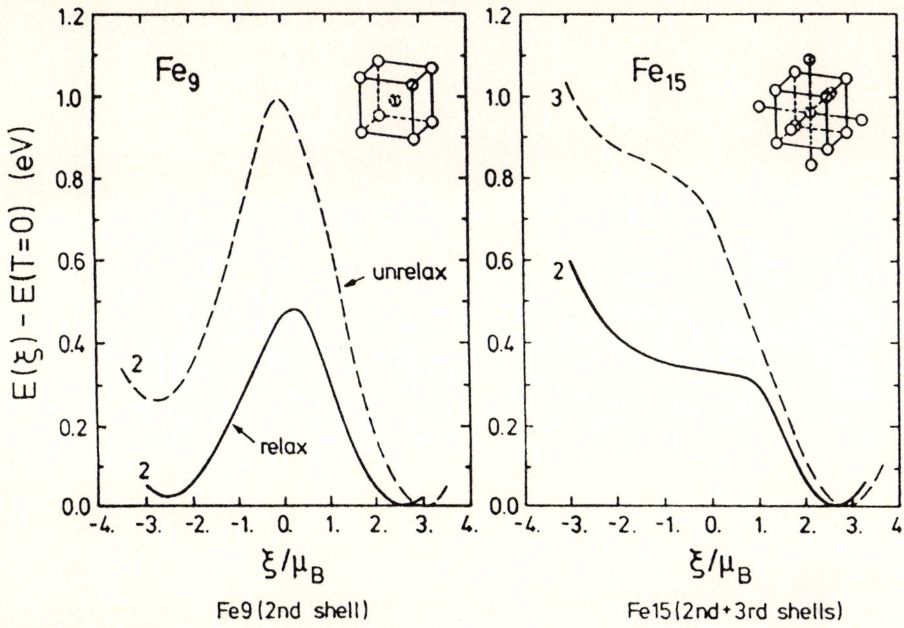

Figure XXV.3. Spin–fluctuation energy $\Delta E(\xi) = E(\xi) - E(T = 0)$ for Fe_n.

References

XXV.1 H. Hasegawa, J. Phys. F **16**, 347 (1986); J. Dorantes–Dávila, G.M. Pastor, and K.H. Bennemann, Solid State Commun. **60**, 465 (1986).

XXV.2 D.M. Cox, D.J. Trevar, R.L. Whetten, E.A. Rohlfing, and A. Kaldor, Phys. Rev. B **32**, 7290 (1985).

XXV.3 W. Keune, R. Halbauer, U. Gonser, J. Lauer, and D.L. Williamson, J. Magn. Magn. Mat., **6**, 192 (1977).

XXV.4 G.M. Pastor, J. Dorantes–Dávila, and K.H. Bennemann, Phys. Rev. B **40**, 7642 (1989).

Part 6

Order – Disorder

XXVI. Structural Disorder at Low Densities

J.R. Martínez [1]*, G. Muñiz* [2]*, and J. Urias* [2]

[1]Facultad de Ciencias, Universidad Autónoma de San Luis Potosí,
 San Luis Potosí, S.L.P. 78000, México
[2]Instituto de Física "Manuel Sandoval Vallarta", Universidad Autónoma
 de San Luis Potosí, San Luis Potosí, S.L.P. 78000, México

The average density of states (DOS) of a tight binding Hamiltonian with off–diagonal disorder is studied. The only source of disorder is the randomness in the atomic–site positions. A scheme to calculate average DOS, based on a consistency relation between the diagonal and off–diagonal elements of the Green function, is presented. The case of exponentially decaying transfer integrals is discussed; results for 1, 2 and 3 dimensions are presented and discussed in detail. For small values of the number density of atoms localized states associated to atomic clusters dominate in the DOS. Complete delocalization of eigenstates is attained in three dimensions at the critical value 0.0662 of the density. At high densities the Matsubara–Toyozawa approximation is recovered.

The absence of any kind of underlying symmetry has been the main impediment to devising a general quantum theory to describe the electronic behavior of amorphous solids [XXVI.1]. The main source of disorder in those systems is the randomness in the atomic–site positions rather than an intrinsic randomness in the atomic features (e.g. the intrasite energy levels). All the standard machinery of solid–state theory based on concepts such as discrete translational symmetry, reciprocal space, Brillouin zones, Bloch states, etc. is no longer useful. Approximation schemes to calculate analytic electronic density of states (DOS) of random atomic arrangements are thus particularly useful [XXVI.2], even if they are based on an independent–electron Hamiltonian. In this contribution an approximation scheme, based on an exact consistency relation between the diagonal and off–diagonal elements of the average Green function in the locator expansion of Matsubara–Toyozawa [XXVI.3], is used to study the average density of electronic states (DOS) of a very simple tight–binding model of structural disorder.

A tight–binding Hamiltonian for N identical single–state atoms distributed randomly in space with an average number density of atoms n is considered. Atoms are placed at random sites R_i and the set $\{R_i\}_n$ defines a particular atomic configuration of number density n and Hamiltonian

$$H = \sum_{i,j} V_{ij}\, a_i^+\, a_j\,. \qquad\qquad (XXVI.1)$$

Springer Proceedings in Physics, Vol. 50   **Magnetic Properties of Low-Dimensional Systems II**
Editors: L.M. Falicov · F. Mejía-Lira · J.L. Morán-López © Springer-Verlag Berlin, Heidelberg 1990

The transfer integrals V_{ij} are

$$V_{ij} = \begin{cases} 0, & \text{if } i = j; \\ e^{-r_{ij}}, & \text{if } i \neq j, \end{cases} \qquad (XXVI.2)$$

where the absolute distance between atoms, $r_{ij} = |R_i - R_j|$, is a random variable. The constant density n shall be taken as the distribution function of r values. We shall refer to the model defined by Eqs. (XXVI.1) and (XXVI.2) as a tight–binding gas (TBG).

A general approximation scheme to calculate average DOS is developed and two limiting cases are considered. The most interesting is the weak coupling limit. The situation corresponds to an atomic array that is poorly intercommunicated and a large fraction of isolated atoms and small clusters are found in the array [XXVI.4]. Below the Anderson critical density n_A the DOS consists of localized states, only. Above but near n_A extended states near $\omega = 0$ are formed, this process is known as Anderson delocalization [XXVI.5]. The DOS contains a fraction of extended states and a fraction of localized states, separated by mobility edges. However, a full description of the evolution of the TBG DOS from low to high densities has not been given. The second case is the strong coupling limit where all the atoms in the array are very well intercommunicated, i.e., the average hopping is large. States in the spectrum appear as a single band of extended states [XXVI.3,XXVI.6].

For a given atomic configuration $\{R_i\}_n$ the Green function can formally be expressed as a perturbative expansion [XXVI.3] in powers of V_{ij}

$$G_{ij} = g_{\circ}\delta_{ij} + g_{\circ}V_{ij}g_{\circ} + \sum_{\nu=1}\sum_{\mu=0}\sum_{l_\nu}{}^{\mu} g_{\circ}V_{il_1}g_{\circ}V_{l_1l_2}g_{\circ}\ldots g_{\circ}V_{l_\nu j}g_{\circ}. \qquad (XXVI.3)$$

The μ index in Eq. (XXVI.3) is the number of intermediate sites for a given order, $\nu+1$, in the perturbation V_{ij}; $\sum{}^{\mu}$ is a multiple summation over all possible choices of μ intermediate sites in the configuration $\{R_i\}_n$ and $g_{\circ}$ is the atomic Green function: $g_{\circ}(z) = 1/z$. Notice that terms with $\mu = 0$ contribute only to the off–diagonal elements of the Green function and the number of intermediate sites cannot be greater than ν since all terms in the expansion Eq. (XXVI.3) are fully connected; some of the l_ν site indices may appear more than once in a term.

The average over all possible configurations $\{R_i\}_n$ of any quantity $f(i,\{l\}_\mu,j)$ can formally be expressed as

$$\left\langle \sum{}^{\mu}_l f(i,\{l\}_\mu,j) \right\rangle = n^\mu \int (D^\mu r)_{ij}\, f(i,\{r\}_\mu,j) \qquad (XXVI.4)$$

where $n^\mu(D^\mu r)_{ij} = (\prod_1^\mu d^m r_k)\rho(i,\{r\}_\mu,j)$ is the probability to find an atom within everyone of the m–dimensional volume elements $d^m r_k$ located at sites r_k, knowing that atoms are found at sites i and j. Here $\rho(i, \{r\}_\mu, j)$ is a $(\mu + 2)$-point distribution function. However this "full" statistical description avoids any possibility of summing up the perturbation series, Eq. (XXVI.3). So, in order to proceed further we shall use the simplest factorization scheme that takes as independent events the atom positions. That is, we take $\rho(i, \{r\}_\mu, j) = n^\mu$, i.e., $(D^\mu r)_{ij} = \prod_1^\mu (d^m r_k)$.

With these definitions the average Green function is written

$$\langle G_{ij} \rangle = g_\circ \, \delta_{ij} + g_\circ \, \Delta_{ij} \, g_\circ \qquad\qquad (XXVI.5)$$

where

$$\Delta_{ij} = V_{ij} + \sum \sum^\nu n^\mu g_\circ^\nu \int (D^\mu r)_{ij} \, V_{il_1} V_{l_1 l_2} \ldots V_{l_\nu j} \qquad\qquad (XXVI.6)$$

and, just as in Eq. (XXVI.3), some of the l_ν site indices may appear more than once in a term. Matsubara and Toyozawa have introduced a set of rules to represent the terms in Eq. (XXVI.5) by diagrams [XXVI.2]; we shall give just a summary description of the rules. A term of order $\nu + 1$ in Eq. (XXVI.5) with μ intermediate sites will be represented by $(\mu + 2)$ $\mu + 1$ dots connected by $\nu + 1$ lines forming a single string that starts at site i and ends up at site (j) i for the (off) diagonal matrix elements. A line connecting site k to site l represents a factor V_{kl}; a factor $g_\circ$ is added every time the string goes through a site and the term is completed with an overall factor of $n^\mu \int (D^\mu r)_{ij}$. To represent everything diagrammatically, let us denote the diagonal element $\langle G_{ii} \rangle \equiv G$ [the same, on the average (!), at any site i] in a diagram by a small black square and for $i \neq j$ let us introduce a fully renormalized jumper J_{ij} connecting sites i and j, with the Green functions removed from the end sites. The full jumper J_{ij} shall be drawn in a diagram as a slashed line.

The important point to our approximation scheme is to notice that G and J_{ij} are not independent quantities but they must satisfy a consistency condition that we have represented diagrammatically in Figure XXVI.1(a). A simple way to see why Figure XXVI.1(a) must hold is that any diagram of order greater than zero (i.e. not a single dot) contributing to G is always at least of order two in the perturbation (i.e. $V_{ii} = 0$) and is just of the form given by the second diagram on the right hand side of Figure XXVI.1(a). Indeed, if one is to draw a diagram starting and finishing at site i the first thing to do is to jump to an intermediate site other than i (call it site l) since $V_{ii} = 0$; once at site l one can decorate it at will and for all (a black square) before going back, in any way (a full jumper), to the starting site i. Finally, site i can also be decorated (a black square). Any term contributing to G is necessarily of this type. To Figure XXVI.1(a) corresponds the expression

$$G = G_\circ + G^2 \, g_\circ \, n \int d^m r \, V(r) \, J(r). \qquad\qquad (XXVI.7)$$

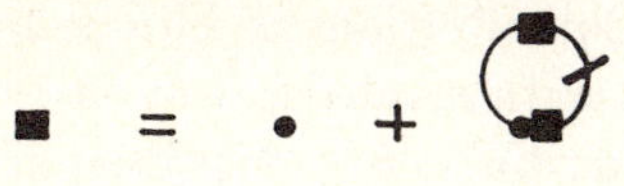

Figure XXVI.1. (a) Consistency condition between the fully renormalized jumper J_{ij} and $\langle G_{ij} \rangle$. (b) Pair approximation for the jumper J_{ij}. (c) String approximation for J_{ij}.

If Eq. (XXVI.7) is iterated repeatedly all kind of Cayley trees are generated, including infinite trees, with an effective transfer given by $g_\circ V_{ij} J_{ij}$; therefore, what Eq. (XXVI.7) makes is to replace the average over configurations for an average over Cayley trees; the same formal average for 1 to 3 dimensions: all the particularities are hidden in the jumper J_{ij}.

So, to calculate G we just need a further equation for the full jumper J_{ij}. Unfortunatly, it is not possible to write an exact expresion for it without introducing an infinite number of multi–line jumper functions. However, approximations for the jumper, suitable to the regime that is being considered, can be deviced. We study here only the two already mentioned limiting cases.

Eq. (XXVI.6) includes a double sum such that for every ν value (i.e., for a given order in the perturbation), μ takes values from 0 to ν. Figure XXVI.2 shows all possible terms of order $\nu + 1 = 5$ contributing to Δ_{ij} (including the decoration of the end sites), classified by their μ (0-4) value. Diagrams with the highest power in n are those with $\mu = \nu$ (4 in Figure XXVI.2). The only possible such diagrams are strings, as shown at the top line of Figure XXVI.2. On the contrary, the terms that are lowest order in n are those with $\mu = 0$ and are independent of the density n. They are diagrams without intermediate sites as the watermelon diagram at the bottom line of Figure XXVI.2.

Let us consider the low density approximation first. As shown at the bottom row of Figure XXVI.2, the watermelon diagrams are zeroth order terms and are thus obligatory contributions in any low density approximation. The equation in Figure XXVI.1(b) gives an approximate full jumper J_{ij} that includes all watermelon diagrams and nothing else: it is a pair approximation for the jumper and corresponds to the expression

$$J_{ij} = V_{ij} + g_\circ^2 \, V_{ij}^2 \, J_{ij}. \qquad (XXVI.8)$$

Eqs.(XXVI.9) and (XXVI.13) can be solved for G

$$G = g_\circ \left(1 \pm \sqrt{1 - 4f(\omega)}\right)/[2f(\omega)], \qquad (XXVI.9)$$

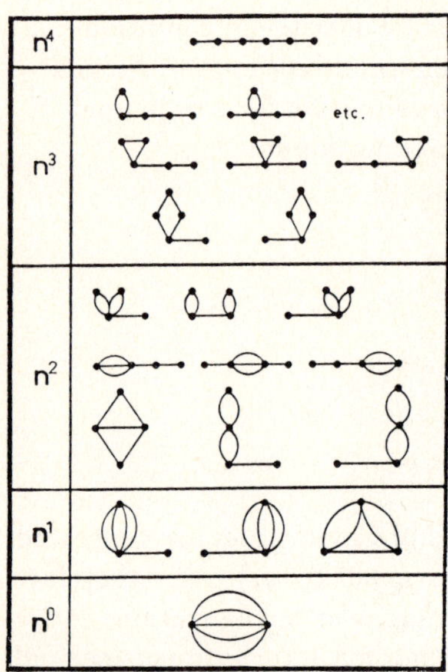

Figure XXVI.2. Contributions of fifth order in V. The diagrams without decoration at the end sites are the contributions to the jumper J_{ij}.

where

$$f(\omega) = n \int d^m r \frac{[g_\circ V(r)]^2}{1 - [g_\circ V(r)]^2}. \qquad (XXVI.10)$$

Before presenting results for the DOS, let us discuss some of the analytical properties of $f(\omega)$ and their consequences for the Green function and the spectrum. For convenience we set the energy scale such that $Max\,|V(r)| = 1$. The function $f(\omega)$ has the following properties:

(i) If $V(r)$ is a continuous function of r, $f(z)$ has a cut on the real axis from $\omega = -1$ to $+1$ and $f(\omega + i0^+)$ is complex valued along the cut $|\omega| < 1$.

(ii) If $V(r)$ is not a continuous function of r, $f(z)$ may either have no cuts or a set of them on the real axis within the energy range $|\omega| < 1$.

(iii) For any $V(r)$, $f(\omega)$ is real valued for $|\omega| > 1$ and the DOS will be non–zero only if

$$f(\omega) > 1/4; \quad |\omega| > 1. \qquad (XXVI.11)$$

The energy values, if any, satisfying Eq. (XXVI.11) correspond to extended eigenstates.

(iv) $f(\omega)$ has maxima at $\omega = \pm 1^\pm$. For a long range interaction, $V(r)$, and low dimensionality $(m = 1)$ $f(\omega) \to \infty$ as $|\omega| \to 1^+$. When this is the case, for any value of the number density n the DOS has tails beyond $\omega = \pm 1$. For short range couplings, as the one we are considering [see Eq. (XXVI.2)], and $m > 1$, $f(\omega)$ is finite at $\omega = \pm 1$. Then a critical value of the number density, n_c, for complete delocalization of states and onset of bandwidth broadening, is given by

234

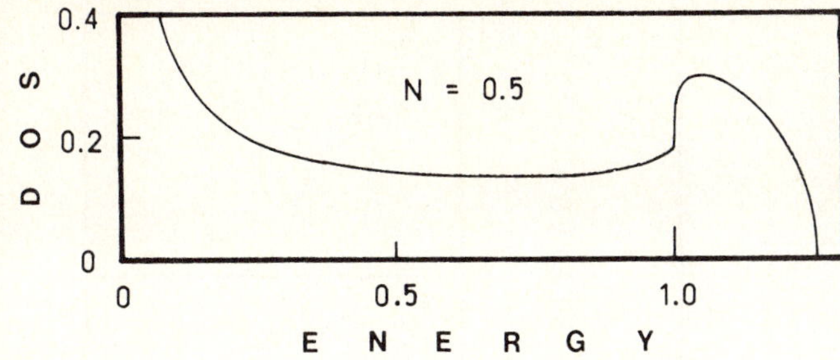

Figure XXVI.3. DOS for a one–dimensional TBG with 0.5 sites per unit sphere. All states are extended, i.e. $n_c = 0$

$$1 = 4n_c \int d^m r \, \frac{V(r)^2}{1 - V(r)^2} . \qquad (XXVI.12)$$

For $m = 1$ the critical density, n_c , for the onset of bandwidth broadening is $n_c = 0$ and the DOS always has two tails beyond the edges at $\omega = \pm 1$. Figure XXVI.3 shows the positive energy part of a one–dimensional DOS for $n = 0.5$. For $m = 2$ and 3 dimensions the critical densities are

$$\Omega_m \, n_c(m) = 1/[m\, \varsigma(m)], \qquad (XXVI.13)$$

where $\varsigma(m)$ is the Riemann's zeta function and Ω_m is the volume of the unit sphere in m dimensions. Numerically,

$$n_c(2) = 0.0967 \qquad \text{and} \qquad n_c(3) = 0.0662 . \qquad (XXVI.14)$$

A sequence of DOS for three values of $N = \Omega_m \, n$, going through $N_c = \Omega_m \, n_c$, are shown for 2 and 3 dimensions in Figs. XXVI.4 and XXVI.5, respectively. Far below N_c the DOS looks very much as an average of pair states [XXVI.6], i.e. the smallest Cayley trees dominate. As N is made bigger the average distance between atoms is shortened, increasing the average pair interaction such that states nearby $\omega = 0$ are pushed towards the band ends at $\omega = \pm 1$ and bigger Cayley trees become important. When the N_c limit is attained all the eigenstates are delocalized and as N becomes greater than N_c states go out the ± 1 limits: the broadening of the DOS bandwidth begins (see Figs. XXVI.4 and XXVI.5). Some of the reported Anderson transition densities, n_A , for the three–dimensional TBG are 5.83×10^{-3} , $(5.1 \pm 1.7) \times 10^{-3}$, 2.7×10^{-3} and 0.0506 as taken from Ref. (XXVI.6). A comparison of the reported n_A values with the $n_c(3)$ in Eq. (XXVI.14) indicates that the former is smaller than the latter by about one order of magnitude (notice that the pair approximation overestimates n_c).

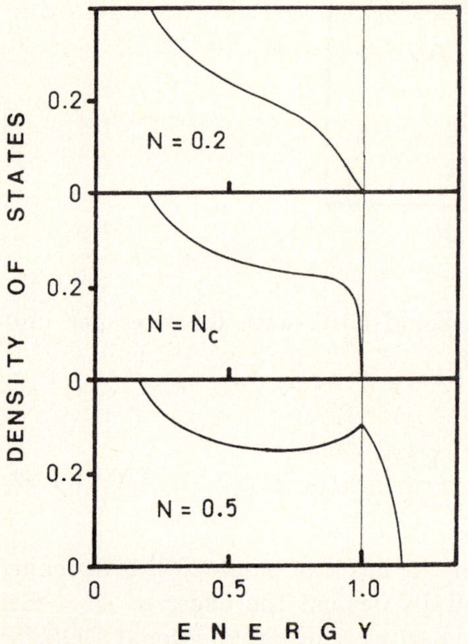

Figure XXVI.4. Sequence of DOS for the two–dimensional TBG for three values of N. At the critical value $N_c = \pi n_c = 0.3039$ complete delocalization is attained.

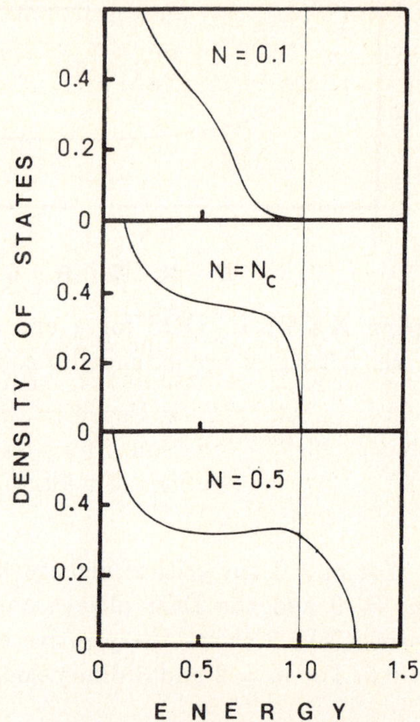

Figure XXVI.5. Sequence of DOS for the three–dimensional TBG for three values of N. At the critical value $N_c = 4\pi n_c/3 = 0.2773$ complete delocalization is attained.

Although the pair approximation Eq. (XXVI.8) is far from leading to a full description of the TBG DOS at low densities, it has provided us with a qualitatively complete picture of the TBG DOS: all states are localized at very low densities, at higher densities extended states appear at the center of the DOS and at the critical density value n_c complete delocalization is attained as the onset for bandwidth broadening. The pair approximation can be improved in a straightforward way by incorporating in Eq. (XXVI.8) contributions of higher order in the number density, n, (see Figure XXVI.2) as in a cluster series expansion for the jumper. The pair approximation, Eq. (XXVI.8), is just the zeroth–order term.

In the high–density limit the most important terms contributing to the full jumper are the string diagrams. Figure XXVI.1(c) shows the equation for J_{ij} equivalent to sum up all string diagrams in all possible forms. With this jumper the Matsubara–Toyozawa approximation [XXVI.3] is recovered. There are no localized states and the DOS has a very long tail at its lower half [XXVI.3, XXVI.6].

We believe that an approximate average DOS that would give a good interpolation from the low– to the high–density regimes of the TBG is possible within the scheme we have presented.

References

XXVI.1 M. F. Thorpe (Editor), *Excitations in Disordered Systems* (Plenum, New York, 1982).

XXVI.2 R. M. Stratt and Bing-Chang Xu, Phys. Rev. Letters **62**, 1675 (1989).

XXVI.3 T. Matsubara and Y. Toyozawa, Progr. Theor. Phys. **26**, 739 (1961).

XXVI.4 I. M. Lifshitz, Usp. Fiz. Nauk. **126**, 41 (1978)

XXVI.5 P. W. Anderson, Phys. Rev. **109**, 1492 (1958).

XXVI.6 M. K. Gibbons, D. E. Logan and P. A. Madden, Phys. Rev. B **38**, 7292 (1988).

Part 7

Magnetism and Superconductivity

XXVII. Antiferromagnetic Order in the CuO$_2$ Planes of High-T$_c$ Superconductors

C.A. Balseiro

Centro Atómico Bariloche, 8400 San Carlos de Bariloche, R.N., Argentina

We study a generalized Hubbard model which includes the Cu and O orbitals in the CuO$_2$ planes of high Tc superconductors. We present some analytic and numerical results obtained by using the slave–boson technique and exact diagonalization of small clusters respectively. We show that in the case of one hole per unit cell, the CuO$_2$ planes are charge transfer insulators. Doping produces a rapid metallization and destroys the antiferromagnetic order. We also present results obtained for the spin and charge dynamics.

The common feature of all high Tc cuprate superconductors are the CuO$_2$ planes which play a crucial role in determining the electronic properties of these materials. In fact, there is now a general consensus that the relevant carriers are localized in these planes [XXVII.1-XXVII.3].

Due to the large electron–electron correlation in the Cu ions [XXVII.4-XXVII.6], the conventional band–structure calculations cannot be used as a good starting point for the understanding of the electronic properties. In fact, this type of approach fails in predicting the true nature of the ground state and excitations of these materials. However, total–energy calculations correctly describe the two following features: the low–energy electronic excitations involves the Cu $3d_{x^2-y^2}$ and the O $2p_x$ and $2p_y$ orbitals and an enhanced covalency in the Cu–O planes as compared to transition metal oxides.

These features support the idea that the simplest model Hamiltonian which correctly describes the electronic structure of the CuO$_2$ planes should include —at least as a starting point— the Cu and O orbitals, a strong hybridization and large on–site correlations as the basic ingredients.

The model contains both Cu $3d_{x^2-y^2}$ and O $2p$ orbitals. In the usual notation the Hamiltonian reads [XXVII.7,XXVII.8]

$$H = \sum_{i\sigma} \epsilon_i n_{i\sigma} + \sum_{<ij>} t_{ij} c^+_{i\sigma} c_{j\sigma} + \sum_i U_i n_{i\uparrow} n_{i\downarrow} + V \sum_{<ij>} n_i n_j, \quad (XXVII.1)$$

where the index i runs over the sites of a CuO$_2$ two–dimensional lattice. The operator $c^+_{i\sigma}$ creates a hole with spin σ at the Cu $3d$ (O $2p$) orbital for i labelling a Cu (O) site. The corresponding parameters $\epsilon_i = \epsilon_p, \epsilon_d$ and $U_i = U_p, U_d$ are

Springer Proceedings in Physics, Vol. 50   **Magnetic Properties of Low-Dimensional Systems II**
Editors: L.M. Falicov · F. Mejía-Lira · J.L. Morán-López © Springer-Verlag Berlin, Heidelberg 1990

the orbital energies and the intra–atomic repulsion respectively, while t_{ij} is the hopping matrix element and V is the interatomic Coulomb repulsion between nearest neighbor sites.

The stoichiometric reference systems such as La_2CuO_4, Ln_2CuO_4 (Ln=Pr, Nd, Sm) or in ideal YBaCuO are insulating or semiconductors. Moreover, La_2CuO_4 and YBaCuO are antiferromagnetic [XXVII.9], while the magnetic properties of Ln_2CuO_4 have not been reported yet. In these materials there is one hole per unit cell in the CuO_2 planes. By replacing trivalent La by divalent ions (Sr, Ba) in the La_2CuO_4, holes are introduced in those planes. Similar "hole doping"is obtained in 1–2–3 compounds by changing the O concentrations.

We first give a "zero order"description of these systems in terms of the parameters of the Hamiltonian (XXVII.1). This is done by taking zero hopping $(t_{ij}=0)$. In the undoped materials the hole is localized in the Cu $3d$ shell $(\epsilon_d < \epsilon_p)$ giving rise to Cu^{2+} and O^{2-}. When holes are added it is energetically favorable to place them in the O $2p$ states. This is a consequence of the large intra–atomic Coulomb repulsion on the Cu $3d$ shell $(U_d > \epsilon_p - \epsilon_d)$. If this is the case, each added hole produces an O^-. With electron doping, some holes are removed from the planes, and then each electron produces a Cu^+. When the hopping term is included, the description becomes much more complicated.

In what follows we present some analytical and numerical results. We present first the results obtained with the slave–boson approach in the saddle point approximation [XXVII.9,XXVII.10].

Assuming a paramagnetic state at $T=0$, for the case of one hole per unit cell, we find in this approximation either a semiconducting or metallic phase. In Fig. XXVII.1 we show the obtained phase diagram in the (U_d, Δ) space, where $\Delta=(\epsilon_p - \epsilon_d + 2V)/2$ is the charge–transfer energy. Note that Δ not only depends on the energy difference between the Cu and O orbitals but also includes the effect of the interatomic repulsion V as suggested by Varma *et*

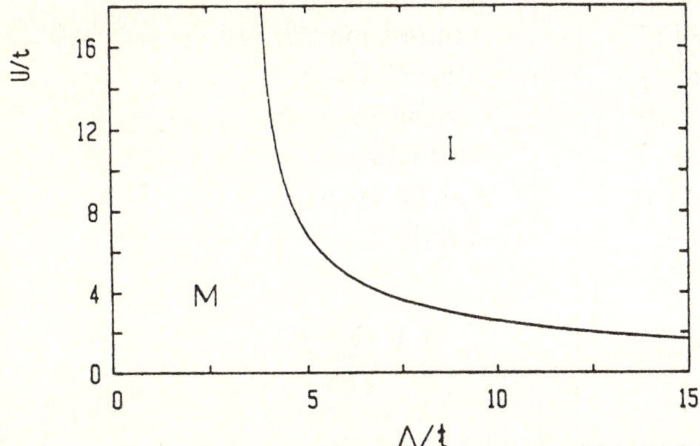

Figure XXVII.1. Metal–insulator phase diagram in the (U,Δ)–plane, $\Delta = \epsilon + V/2$.

al. [XXVII.11]. This phase diagram resembles the Zaanen–Sawatzky–Allen phase diagram [XXVII.12]. For large U_d, by increasing Δ one reaches a metal–insulator boundary at $\Delta^c \simeq 2t + 2t^2/U_d$. The semiconducting state is of charge transfer type. In the large Δ limit, the boundary is at $U_d^{(c)} \simeq 8t^2/\Delta$. This limit can be viewed as a Mott transition in the narrow bonding band. For $\epsilon_p > \epsilon_d$, the phase boundary is independent of U_p. This seems to be a quite general phase diagram, since we obtain very similar boundaries for different lattice structures if Δ is replaced by $\Delta = (\epsilon_p - \epsilon_d + zV)/2$, where z is the oxygen coordination number of the lattice. Within the same framework we study the antiferromagnetic instabilty. We assume an antiferromagnetic structure in the Cu sublattice. By studying the divergence of the staggered susceptibility, we obtain the phase diagram of Fig. XXVII.2. This phase diagram is restricted to $\epsilon = (\epsilon_p - \epsilon_d)/2 > 0.5$. For ϵ small (or negative) the holes tend to localize on O sites and consequently, the proposed magnetic structure becomes unstable. For $\epsilon \geq 0.5$, the holes are localized mainly on the Cu and the phase diagram is not very sensitive to U_p. In summary, we have shown that for large U_d and for Δ larger than a critical value, the stoichiometric systems are charge–transfer insulators.

For a reasonable set of parameters obtained for La_2CuO_4 [XXVII.13] ($t \sim$ 1.5 eV, $U_d \sim 6t$, $U_p \sim 3t$, $\epsilon \sim 2t$, $V \sim 1.5t$) this system lies in the region of charge–transfer semiconductors. However, for this set of parameters the system is close to the metal–insulator boundary and with a change in the parameters within physically reasonable limits one could obtain either a metal or a semiconductor with a small gap. In agreement with experimental results, systems close to stoichiometry are antiferromagnetic but the long–range antiferromagnetic order is rapidly destroyed with doping.

It is interesting to compare these results with exact results obtained numerically for a small cluster. We consider a cluster of 12 atoms (four unit cells)

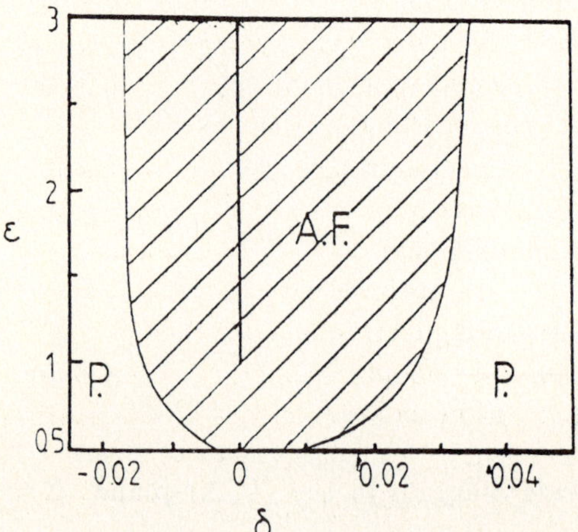

Figure XXVII.2. Antiferromagnetic phase diagram in the (U,δ)–plane, $\delta = (n-1)$.

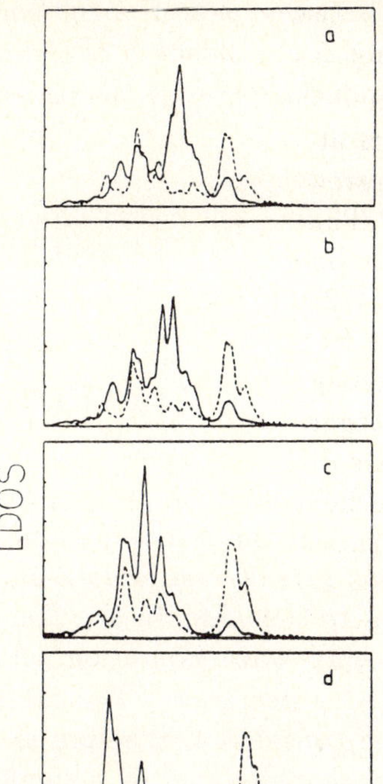

Figure XXVII.3 One–particle spectral densities for the Cu (dashed line) and oxygen (continuous line). The parameters are $U_d/t = 8$, $U_p/t = 4$, $V/t = 1$, and (a) $\epsilon/t = 0.25$, (b) $\epsilon/t = 0.5$, (c) $\epsilon/t = 1$, (d) $\epsilon/t = 2$.

and calculate expectation values of different quantities as well as frequency–dependent correlation functions [XXVII.14].

We first calculate one particle spectral densities for the Cu and O ions. The results obtained for stoichiometric systems (four holes in the cluster) are shown in Fig. XXVII.3, for $U_d = 2t$, $U_p = 4t$, $V = t$ and different values of ϵ.

For large ϵ the system is clearly a semiconductor with a gap which is of the order of 2ϵ, i.e. a charge–transfer insulator. As ϵ decreases, the gap decreases and eventually disappears although with this small–cluster calculation, it is not possible to give correctly the detailed behavior of the gap.

Moreover, at zero temperature, in the thermodynamic limit it is possible that even for small ϵ the system has a non–zero gap, due to the long–range magnetic order. This gap, however, should disappear at the Néel temperature in contrast with the results obtained for large or intermediate values of ϵ, where the gap is almost independent of the spin configuration. With this in mind, we can say that the behavior obtained for the one–particle spectral densities, supports the idea of a metal–insulator transition as ϵ decreases, in agreement with the phase diagram of Fig. XXVII.1.

Now, we present results corresponding to the magnetic properties of these clusters. The equal–time correlation function indicates, for the undoped system, a strong antiferromagnetic (AF) correlation between nearest–neighbor (n.n.)(ferromagnetic between second n.n.) copper spins. The strength of these correlations, which are consistent with an AF ordering of the Cu spins, decreases as ϵ decreases, again in agreement with analytical results.

In searching for the spin and charge excitations we define

$$S_q^z = \sum_j e^{iqR_j} S_j^z, \qquad\qquad (XXVII.2)$$

and

$$n_q = \sum_j e^{iqR_j} n_j, \qquad\qquad (XXVII.3)$$

where j runs over all Cu sites. We calculate $\langle S_q^z, S_{-q}^z \rangle_\omega$ and $\langle n_q, n_{-q} \rangle_\omega$ for wavevectors $q_1 = (\pi/a, \pi/a)$ and $q_2 = (0, \pi/a)$, where a is the lattice parameter. For both values of q, the spin correlation function presents an intense line at a frequency ω_q and low density satellites at higher frequencies. In the thermodynamic limit, the intense line corresponds to a spin–wave excitation. As ϵ decreases, all lines lose intensity indicating that $\langle S_j^{z\,2} \rangle$ decreases. The ratio ω_{q1}/ω_{q2} makes evident that for the parameter range considered, the spin excitations cannot be described by a Heisenberg model characterized by a single parameter J (for our cluster a Heisenberg model gives $\omega_{q1}/\omega_{q2}=2$). Although it is clear that for small ϵ a canonical transformation cannot be performed to obtain a spin Hamiltonian, for intermediate values of ϵ, where we expect the perturbations to give a reasonable description of the spin excitations, we found deviations from the Heisenberg result. For example, for $\epsilon = t$, corresponding to a charge–transfer energy $2\epsilon + V = 3t$, we obtain $\omega_{q1}/\omega_{q2}=2.7$. It has been shown that in these systems a four spin exchange can be important [XXVII.15]. Our results are consistent with the existence of such a term.

We consider the effective exchange Hamiltonian

$$H_{eff} = -J \sum_{i<j} P_{ij} - K \sum_{i<j<k<l} (P_{ijkl} + P_{ijkl}^{-1}), \qquad\qquad (XXVII.4)$$

where P_{ij} and P_{ijkl} are the two– and four–spin cyclic permutation operators [XXVII.16]. By comparing the results of the spin dynamics of the complete Hamiltonian (XXVII.1), we obtained the values of coupling constants J and K shown in Fig. XXVII.4.

In summary, we have shown that: *(i)* stoichiometric systems with large or intermediate values of $\epsilon_p - \epsilon_d$ are charge–transfer insulators. As $\epsilon_p - \epsilon_d$ decreases the system undergoes a meta-insulator transition as suggested by the impurity calculations of Zaanen et al.[XXVII.12]. *(ii)* For one hole per unit cell, the CuO_2 planes are antiferromagnetic. However, a small doping destroys the antiferromagnetic long range order. *(iii)* In these systems the spin dynamics

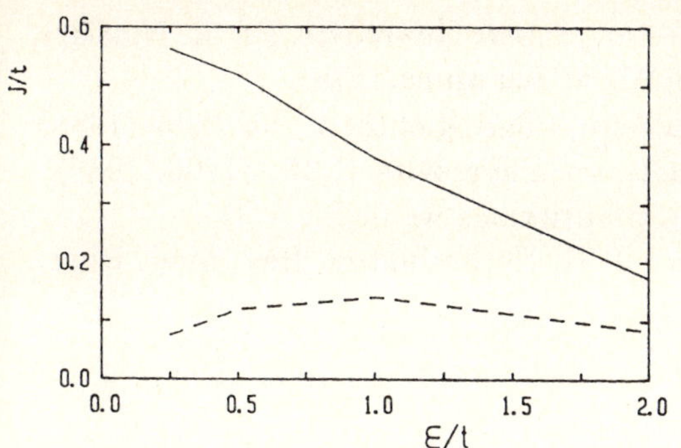

Figure XXVII.4. The coupling constants J (continuous line) and K (dashed line) as a function of ϵ.

can be described by a Heisenberg Hamiltonian only for very large values of ϵ. For values of ϵ usually accepted, a four spin interaction term should probably be considered.

References

XXVII.1 See for example: *High Temperature Superconductors and Materials and Mechanisms of Superconductivity*, Eds. J. Muller and J.L. Olsen, North Holland, 1988.

XXVII.2 F. C. Zhang and T. M. Rice, Phys. Rev. B **37**, 3759 (1988).

XXVII.3 V. J. Emery, Phys. Rev. Lett., **58**, 2794 (1987).

XXVII.4 A. Fujimori, E. Takayama–Muromachi, Y. Uchida, and B. Okai, Phys. Rev. B **25**, 8814 (1987).

XXVII.5 J. C. Fuggle, P.J.W. Weijs, R. Schoorl, G.A. Sawatsky, J. Fink, N. Nücker, P.J. Durham, and W.M. Temmerman, Phys. Rev. B **37**, 123 (1988).

XXVII.6 Christensen (to be published); F. Mila, Phys. Rev. B **38**, 11358 (1988).

XXVII.7 C.A. Balseiro, A.G. Rojo, E.R. Gagliano, and B. Alascio, Phys. Rev. B **38**, 9315 (1988). Also in Ref. 1, p. 1223.

XXVII.8 J. Hirsch, S. Tang, E. Loh Jr., and D. J. Scalapino, Phys. Rev. Lett. **60**, 1668 (1988).

XXVII.9 G. Kotliar and A. Ruckenstein, Phys. Rev. Lett., **57**, 1362 (1986).

XXVII.10 G. Kotliar, P. A. Lee, and N. Read, in Ref. 1, p.538, P. J. Hirschfeld, G. Kotliar, and A.E. Ruckenstein, (APS March Meeting, New Orleans, 1988) unpublished.

XXVII.11 C. M. Varma, S. Schmitt–Rink, and E. Abrahams, Solid State Commun., **62**, 681 (1987).

XXVII.12 J. Zaanen, G. A. Sawatzky, and J. W. Allen, Phys. Rev. Lett., **55**, 418 (1985).

XXVII.13 M. Schlüter, in *High Temperature Superconductivity*, Ed. R. Nicolsky, R. Barrio and R. Escudero (World Scientific, 1988).

XXVII.14 E. R. Gagliano and C. A. Balseiro, Phys. Rev. Lett., **55**, 2999 (1988); E. R. Gagliano and C. A. Balseiro, Phys. Rev. B **38**, 11766 (1988).

XXVII.15 M. Roger and J. M. Delrieu, preprint.

XXVII.16 M. Roger, J. M. Delrieu, and J. H. Hetherington, Rev. Mod. Phys., **55**, 1 (1983).

Index of Contributors

Springer Proceedings in Physics

Managing Editor: H. K. V. Lotsch

SPP
Band 1–29
6/89

Magnetism and Superconductivity